northrop **flying wings**

by edward t. maloney

edited by
donald w. thorpe

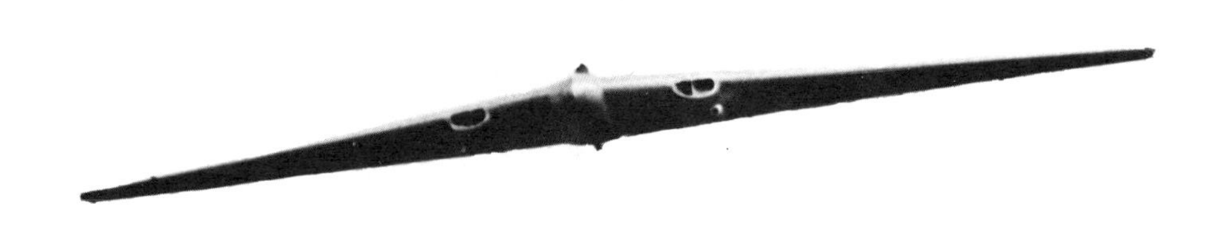

LIBRARY OF CONGRESS CATALOG CARD NUMBER 74-25469
ISBN 0–915464–00–4

PRINTED IN THE UNITED STATES OF AMERICA

WORLD WAR II PUBLICATIONS
Post Office Drawer 278
Corona Del Mar,Calif.92625

Revised Edition 1983
4th Printing

DEDICATION

This book is humbly dedicated to the genius of John K. Northrop and his family of co-workers who built and perfected America's most advanced aircraft desigh of the 20th century.

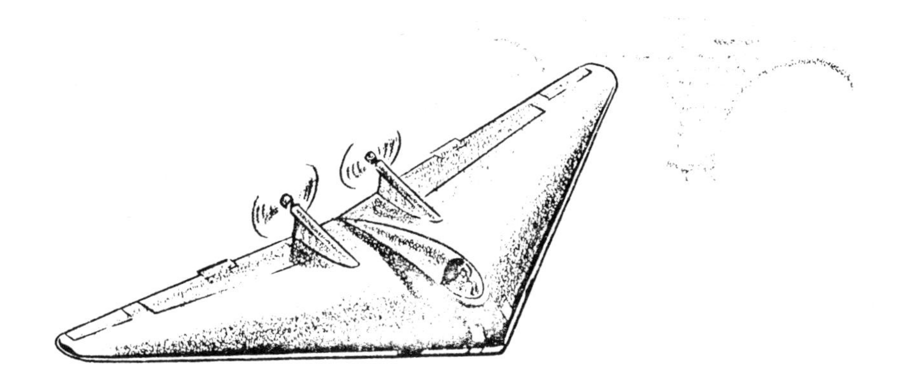

The author wishes to express his appreciation to the following persons and organizations for their help in the preparation of this book:

Northrop Company
Air Force Museum
A. U. Schmidt
Gordon S. Williams
Robert Bewinter
Dustin Carter
Sun Graphics
 Paul M. Freeman, Jr.
 Joseph R. (Chip) Miller
 Larry Cheek
Paul Bickle
Moye Stephens

Cover: The flying wing concept has intrigued aircraft designers since the time of the Wright brothers. The final goal was the elimination of the fuselage and tail so that the airplane became a total lifting body which could fly further, faster and more economically. The XB-35 is shown here flying majestically over the desert on a test flight from Muroc Air Field.

TABLE OF CONTENTS

		MODEL	POPULAR NAME	PAGE
1929	Northrop	Twin Boom Wing		1
1940	Northrop	N1M	"Jeep"	3
1942	Northrop	N9M	"Flying Wing"	6
		N9MB		6
1943	Northrop	XP-56	"Dumbo"	10
1944	Northrop	MX-324	"Rocket Wing"	12
1944	Northrop	MX-543	"Flying Bat"	15
1944	Northrop	XP-79		16
1945	Northrop	XP-79B	"Flying Ram"	17
1944	Northrop	JB-1	"Jet Bomb"	15
1944	Northrop	JB-10	"Buzz Bomb"	15
1946	Northrop	XB-35	"Flying Wing"	19
		YB-35	"Flying Wing"	26
1947	Northrop	YB-49	"Flying Wing"	26
1950	Northrop	YRB-49A	"Flying Wing"	26
	Flight Control Operations			37
1950	Northrop	X-4	"Bantam"	39
	Northrop Test Pilots			41
	Aircraft Specifications			43
	Three Views			44
	Military Serial Number Allocation			44 - 53
	Epilogue			55

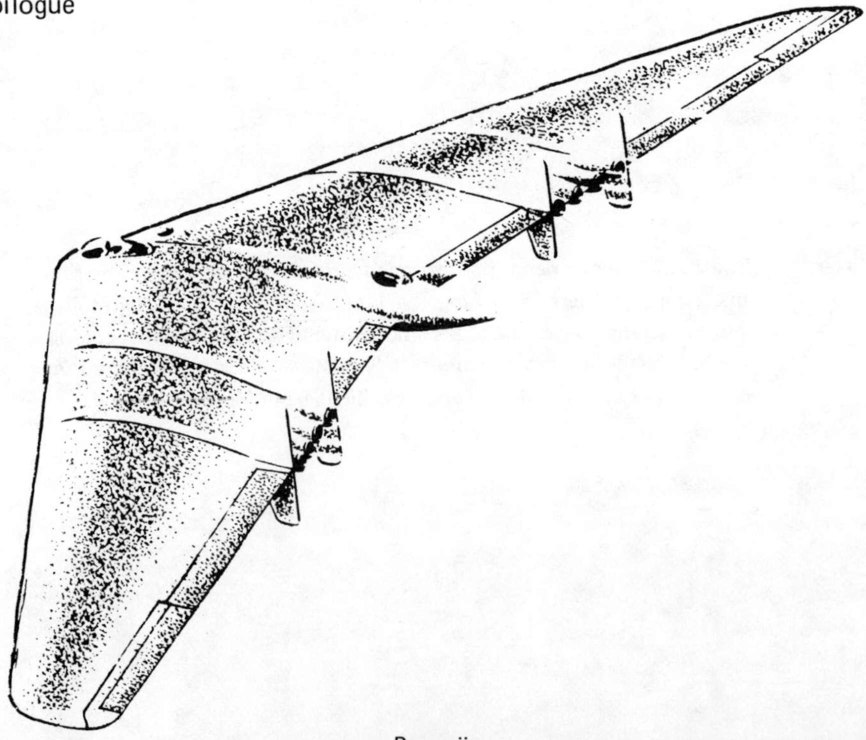

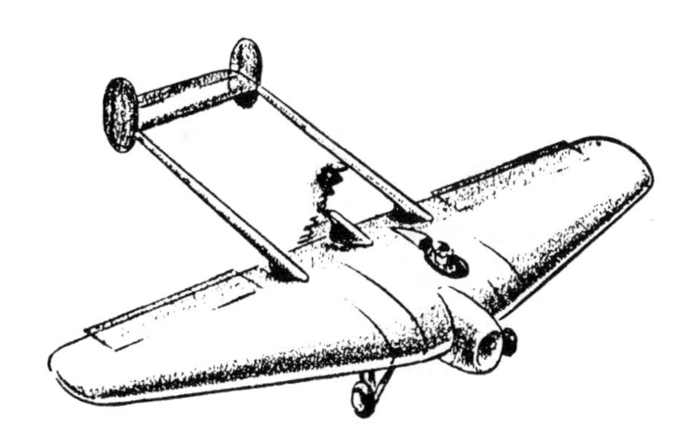

A design as aeronautically clean as the Flying Wing has a big advantage over conventional aircraft design. This advantage is that drag has been reduced to a minimum. And as a result of this minimum drag, the performance of the Flying Wing is unequaled in speed, range and operating economy.

For many years airplane designers have realized that by reducing drag -- the drag caused by the shape or form of an aircraft -- its performance could be greatly increased. Early steps in this direction were changes from biplane to monoplane design; the elimination of external wing struts and flying wires; and the incorporation of retractable landing gears. However, in spite of all these advancements, the average conventional airplane of today still has two to four times the drag of a flying wing. So in order to reduce drag to its absolute minimum, aircraft designers took the drastic step of eliminating the fuselage and tail and putting the pilot, the power plant and payload in the wing envelope.

The chief exponent of Flying Wing design in the United States was John K. Northrop. John Northrop was born in Newark, New Jersey in 1895. When he was nine years old his family moved to Santa Barbara, California. His great interest in mechanics curtailed his formal education, and, after brief periods as garage mechanic, carpenter and draftsman, he went to work in 1916 for the Loughead Brothers, who were then engaged in building twin-engine flying boats in Santa Barbara. Except for a stint with the Army Signal Corps during World War I, Northrop stayed with the Lougheads until 1923. In that year, he joined the Douglas Aircraft Company, which was then preparing the famous Around-the-World Cruisers. He stayed for four years working as draftsman, designer and project engineer on many early Douglas planes.

In January, 1927, John Northrop and three other men formed the Lockheed Aircraft Company. It was at this time that he designed the famous Lockheed Vega — a high wing, cantilever monocoque airplane which was far ahead of its time in design. It was widely used by many of the top flyers such as Amelia Earheart, G. Hubert Wilkins, Frank Hawks, and the noted Wiley Post. Many speed records were set by the Lockheed Vega aircraft.

In the summer of 1928, John Northrop began his research of flying wing aircraft. He formed a small engineering group known as the Avion Company. He built and test flew one of the first semi-flying wing aircraft in the country. Although it did carry the pilot and engine power plant in the wing, it was not a true flying wing because of its small tail carried on twin booms. This plane made numerous flights in 1929 and 1930 and recorded much valuable research data. Unfortunately, the depression of the early 1930's caused further research to be abandoned.

This 1929 Flying Wing design was not a pure flying wing since it had twin booms which carried the tail. This tractor powered model was powered by a Menasco A-4, four cylinder air cooled engine of 90 h.p.

Rare in-flight view of 1929 Flying Wing over Harpers Dry Lake, California. This model did not have all the factors of stability necessary for the elimination of the tail, yet this was a radical design for 1929.

The original tractor powered 1929 design was later re-engined as a pusher. Powered by a Cirrus engine which was barried internally in the wing, this model gave invaluable research data for later wing designs.

As years passed and further advancements were made, John Northrop became recognized as the genius behind the Flying Wing design and the leading exponent of flying wing design in the United States. In Germany the flying-wing-design concept advantages were also shared by Alexander Lippisch and the Horton Brothers and in England by Hill. All were designers, who built and flew successful Flying Wing aircraft. Because it had no fuselage, even the Wright Bi-plane can be considered in this category in spite of its multiplicity of wing struts and wires. The Wrights's plane carried the power plant and pilot directly on the wing system and was controlled longitudinally by a front elevator which was, in fact, a lifting wing surface.

In 1923 John Northrop became interested in the flying wing design after a discussion with Tony Stadlman, who had worked with Northrop in 1917 at the Lockheed factory, then located in Santa Barbara. Stadlman was plant manager of the Douglas Company in Los Angeles at this time and young Northrop was an engineer. Stadlman believed the ideal aircraft of the future was in the true "flying wing" design. From that day on John Northrop's goal was to design and perfect the Flying Wing.

During the early stages fo design development, Northrop held discussions with the late Dr. Theodore von Karman of the California Institute of Technology. Northrop later hired Karman's assistant, Dr. William Sears, to become his Chief Aerodynamicist. Walt Cerny, who came to work for Northrop in 1929, was made assistant Chief of Design. These men had direct supervision of the Flying Wing from its inception. The encouragement and assistance of these men, plus the enthusiastic collaboration of the many Northrop employees, enabled this amazing design to become a reality.

The pressure of designing and building conventional aircraft prevented complete concentration on the solution of the Flying Wing design, and it was not until later that they were able to focus their efforts on Northrop's dream of a true flying wing.

Northrop's first semi-flying wing plane was flown in 1928. It did make use of external control surfaces and carried outrigger twin booms -- quite a radical aircraft for its time. The fuselage was completely eliminated, and the pilot was housed in the wing along with the power plant.

The big problem overcome in designing the flying wing revolved around the buried engine concept. The entire engine propeller shaft had to be buried within the wing foil. This increased the problems of cooling the engine and of turning the long drive shaft to the propellers.

The dimensions of the first Northrop semi-flying wing were: It had a wing span of 30 feet, 6 inches, its length was 20 feet, and its overall height was five feet. Wing area was 184 square feet and aspect ratio was 5.12 to 1. Landing gear was of the tricycle type with the main wheels forward and a tread of 9 feet between the main wheels. Power was supplied by a single four-cylinder 90 horsepower Menasco engine for the tractor powered version. This aircarft later was powered by a Cirrus Engine when it was rebuilt as a pusher-design aircraft.

N1M "Jeep"

In July, 1939, John Northrop began engineering tests for a new flying wing design. This design was known as the N1M "Jeep". The first flight tests were conducted at Muroc Dry Lake in July 1940. Test pilot for the first flight was Vance Breese. Early tests showed the plane to be satisfactory in stability and control. Shortly thereafter, the flight test program was turned over to Mr. Moye W. Stephens, Northrop Test Pilot and Secretary to the Northrop Corporation. During 1940 and 1941, over 200 flights were made in this aircraft to gather further data. Control of the aircraft was achieved through the use of a system of elevons and wing tip rudders. The elevons served in tailless type aircraft both as elevators and ailerons. Rudder action was provided by control surfaces incorporated in the drooping wing tips.

This wing had a wing span of 38 feet and a length of 17 feet, an overall height of 5 feet, a wing area of 300 square feet, and an aspect ratio was 4.75 to 1. It, too, had a tricycle landing gear. It originally carried two four-cylinder Lycoming engines of 65 horsepower each, which were later replaced by two six-cylinder Franklin air-cooled engines of 120 horsepower each. The drooping wing tips of the N1M were later eliminated.

The N1M "Jeep" was truly the first American flying wing aircraft. Today it resides in the National Air and Space Museum Storage facility at Silver Hill, Maryland, a short distance outside of Washington, D. C. It is hoped that this aircraft is one that will be displayed in the new Air and Space Building when it is completed on the Mall in Washington, D.C.

The World's first successful pure flying wing design was the N1M "Jeep" of 1940. Over two hundred successful flights were made with this model. The aircraft was initially painted a bright yellow.

Three-quarter rear view of N1M gave the appearance of a pre-historic bird. It was first powered by two 90 h.p. Lycoming engines and later by two 120 h.p. Franklin engines.

The N1M "Jeep" was truly the first American Flying Wing Aircraft. Today it resides in the National Air and Space Museum Storage Facility at Silver Hill, Maryland.

A N1M at rest at Harpers Dry Lake. The early success of the N1M paved the way for later N9M and XB-35 flying wings.

Many Northrop employees will recapture with nostalgia this scene of the N1M and some of the "Northrop Family" who made the dream of a Flying Wing a reality. John Northrop is just to the left of the nose gear. Moye Stephens is front center. He served as company evaluation test pilot after Vance Breece first test flew the aircraft.

The N1M is shown here with straight wing. Earlier designed drooped wing tips were deleted when they were found to be unnecessary and Northrop designers learned more facts about the pure Flying Wing design.

N9M Wing Series

Upon the urging of General H. H. "Hap" Arnold, Air Force Chief, John Northrop began the investigation and application of the all-wing principle to heavy bombardment-type aircraft. To serve this end, four 60 foot Northrop N9M Flying Wing aircraft were constructed in 1942. They were built of plywood and metal to test the Flying Wing design, to train pilots in Flying Wing airplanes, and to see exactly what could be expected of a large version of this type aircraft. In reality, the N9M was a miniature XB-35 Flying Wing Bomber. It embodied all the improvements learned in the N1M "Jeep".

The N9M's were flown hundreds of hours at Muroc Dry Lake. These flights served the purpose of proving the flight characteristics of flying wing airplanes and of indoctrinating more pilots in the use of this new type of aircraft.

The N9M was originally powered by a pair of six cylinder air-cooled Menasco engines which drove extension shafts to the pusher propellers via a fluid-drive coupler. It had a tricycle landing gear. A later version of the Northrop N9M known as the N9M-B featured two air-cooled Franklin engines of about 400 horsepower each.

All the design details which were worked out in the N9M model were incorporated into the larger XB-35 Flying Wing. After completion of the flight test program, the single-place N9M's were relegated to pilot training. They were used for a number of years to train Air Force pilots in the handling of flying wing aircraft.

The N9M in flight over Southern California. This model carries two Menasco C65-4 engines of 290 h.p. each. The test pilot aboard was John Meyers.

Number one prototype N9M.

John Meyers test flys the No. 1 prototype N9M over Southern California. Provisions were made to carry pilot and passenger in the N9M. Passenger Bill Sears, Northrop engineer, was flown in the N9M during many test flights.

Although over thirty years old, the wing design is as modern today as it was then. It is ageless.

N9M in its hanger at Hawthorne, California. Engine accessory panels are shown removed. Large bell housing is fluid drive coupler to engine which turned the propeller shafts, driving two Hamilton Standard pusher propellers. Note the rear outrigger tail wheel. N9M's were of mixed wood and metal construction. The center section was built up of welded steel tubing. The covering was of wood and metal panels. The outer wing panels which were butted on, were constructed of wood with metal wing slots and wing tips.

A Northrop N9M-B on the apron at Edwards AFB, California. Tricycle landing gear enabled easy take off and landing. A clear bubble canopy provided extremely good visibility.

Top view shows a clear view of the N9M-B in flight over the Mojave Desert. This design paved the way for the larger B-35's and B-49 Flying Wing Jet Bombers. Note the elevons at the extreme end of the wing. The aircraft was painted silver.

Rare view of three Flying Wings in formation. From left to right are the N9M-B, powered by two Franklin air cooled engines, and two N9M's, powered by two Menasco engines each. Note the different locations of the U.S. national insignia. The center N9M is improperly marked. Aircraft at this time were painted silver overall and were in use as pilot trainers at Edwards AFB.

XP-56

One of the most unusual Northrop designs was the XP-56 fighter. This 1943 design had a very short fuselage mounted on a swept wing. Originally a Pratt-Whitney experimental X-1800 liquid cooled engine was to have been mounted when the engineering studies were first ordered in June of 1940, but two prototypes purchased were changed to carry the Pratt-Whitney air-cooled R-2800 twin Wasp radial engine. The XP-56 was designed to carry two 20 m.m. cannons and four 50 caliber machine guns. The number one prototype made its first test flight on September 6, 1943 and was flown by John Meyers. It overturned upon landing due to a malfunction of the landing gear shimmy dampener. The number two prototype was first flown by test pilot Harry Crosby on March 23, 1944. It utilized all magnesium construction. The XP-56 engine was cooled through two wing root intakes, which turned two counter rotating propellers mounted behind the dorsal and ventral fins. It had a tricycle landing gear like other Northrop aircraft and a unique wing which featured drooping wing tips with air operated bellow rudders. Control surfaces were elevons of the type later seen on other flying wings.

The XP-56 was the first Air Force pusher to have counter rotating propellers and the first to have an air cooled engine completely submerged within its fuselage.

The front view picture of the XP-56 gave it an unusual appearance. The XP-56 was an extremely advanced pusher type plane for its day. Details of its existence were not released until 1945.

Engine run up of the XP-56 in early 1943 at the Hawthrone factory. The upper vertical fin was deleted on the prototype aircraft; it was a later modification. The XP-56 was an extremely revolutionary design being an advanced pusher type tailless fighter plane for the Army Air Force. It was the first all magnesium airplane and the first to be of all welded structure.

The Northrop XP-56 utilized a tricycle landing gear. This semi-Flying Wing design, a tailless pursuit plane, carried a crew of one and mounted a pusher type R-2800 air cooled engine. Elevons and lateral controls were mounted on the wing trailing edge. Note the bellow type inlets on the wing tips.

The XP-56 was the first Air Force pusher to have counter rotating propellers and the first to have an air cooled engine completely submerged within its fuselage.

Northrop MX-324

In 1944 John Northrop designed his famous "Rocket Wing" which recieved the designation of MX-324. Northrop's Rocket Wing was America's first military rocket plane. It was completed under great secrecy, and the world did not learn of it until early 1947. Its first flight -- the first military rocket plane flight in the United States -- was made in the presence of a few Northrop officials and United States Air Force personnel at Harpers Dry Lake near Barstow, California on July 5, 1944.

The rocket wing was a Flying Wing of less than 30 feet wing span which employed the new principles of rocket propulsion, and, for the first time, featured a prone pilot. The prone cockpit design enabled the pilot to lie flat on his stomach in order to withstand higher acceleration and "g" forces. It also made a thinner air foil possible. The rocket wing was powered by an Aero Jet XCAL-200 rocket motor. Nitric acid was used as fuel. Some research models were designed to take off and land on skids, although the main prototype aircraft used a fixed tricycle landing gear. The noted pre-war air race pilot, Harry Crosby, was test pilot of the rocket wing during its research program. Although the United States had no jet fighters ready in time for combat during World War II, several were in the design stage.

While the XP-79 was going through drawing board and fabrication phases, tests had begun with the wooden flying scale models which had been assigned the designation MX-324 and, which were referred to as "Project 12" for security reasons. The first of the trio of MX-324 gliders was fitted with skids and was intended to be towed at high speed behind a car, but, owing to the weight of the glider, the car was unable to attain a sufficient speed for the MX-324 to become airborne. The second MX-324 was fitted with a four-wheel detachable trolley, but

The 1944 Northrop MX-324 Rocket wing was America's first military rocket airplane. Completed in great secrecy, the world did not learn of its existence until early 1947. The first rocket flown in the United States was made by this airplane at Harpers Dry Lake near Barstow, California on July 5, 1944.

this also proved unsatisfactory. The third had a fixed tricycle undercarriage which spoiled the lines of the aircraft but at least enabled it to be towed from the ground. The first gliding flight was made on October 2, 1943 with test pilot John Myers at the controls.

From the outset, it had been planned to apply rocket power to the MX-324, and Aerojet had been developing a small acid-aniline motor, the XCAL which produced 200 pounds thrust. It was built specifically for installation in this glider. At Aerojet Engineering plant in Azusa, California this rocket motor was known as "Project X."

The XCAL-200 rocket motor had a cast aluminum combustion chamber and was restartable in the air. Its chief drawback was in its heavy weight of 427 pounds. Like the larger, more sophisticated motor that Aerojet was struggling to produce for the Northrop XP-79, the XCAL had its share of problems, and its delivery was progressibely delayed. During this time gliding trials with the Northrop MX-324 continued.

At this time it was believed that flight control of the XP-79 would improve if fixed slots were installed in the wing. These fixed slots were installed for one flight test, to evaluate their efficiency during flight conditions. This flight nearly ended in tragedy, although the accident was in no way due to the wing slots.

On this occasion, the MX-324 was towed aloft to an altitude of 10,000 feet, and, after casting off the tow line, the pilot accidently released the escape hatches while banking steeply. The pilot nearly fell from the aircraft. His instinctive grasp on the crossbar control produced an incredible wingover. When things calmed down, the pilot found himself hanging upside down from his harness, his head and trunk suspended beneath the aircraft which had entered a steady inverted glide. The minor movements of the crossbar, of which he was able to make seemed to produce little effect. Fortunately, he eventually succeeded in pulling himself onto the leading edge of the center section where he sat badly shaken, while he checked his chute harness. He then slid off the wing, after unhooking his restraining harness and made a normal parachute desent, leaving the aircraft unmanned.

The MX-324, apparently undisturbed by the change in the center of gravity by the pilots departure, continued its inverted glide in a wide circle. It eventually landed upside down a short distance from the point where it had originally taken off. It had suffered only minor damage which was quickly repaired.

The XCAL-200 rocket motor had attained sufficient reliability by the spring of 1944 to permit its installation in the Northrup MX-324. The rocket motor was housed in the wing sectic immediately aft of the pilots cockpit.

The MX-324 rocket wing was powered by an Aerojet XCAL-200 rocket motor. The front view shows the good position of the pilot in the cockpit. The prone cockpit layout enabled the pilot to lie flat on his stomach to withstand higher acceleration. It also made the use of a thin air foil possible. The fixed landing gear was utilized to speed the building process of the aircraft and the four externally braced rudder wires to the rudder. This aircraft is now held in storage by the National Air and Space Nuseum. This was America's first rocket-powered aircraft.

Northrop Buzz Bombs

During the late stages of World War II, Northrop Aircraft Company built, under confidential and secret contracts, several jet, flying wing bombs. The first of these prototype flying wing bombs was the Northrop JB-1. This was ground-launched flying bomb which had a pre-set control which could be ground launched into enemy territory and crash-dived into the target. It carried two 2,000 pound bombs and had a range of 200 miles. It was powered by two turbo-jet engines. Take-off was accomplished from a 400 foot level ramp with five 10,000 pound thrust rockets. Only one prototype JB-1 was ever built and tested. An experimental full-size glider version of the JB-1 was built to test the flight characteristics of the design. It was the same size and dimensions as the JB-1. It was also fitted with a pilot's cockpit. This craft was towed aloft by a parent aircraft and released upon reaching altitude to glide down and gather research data on stability, control and flight characteristics of the design.

Northrop also built another jet flying wing bomb designated JB-10. The JB-10 had a top speed of 400 miles per hour. It was built with an American version of the German "V-1" pulse jet engine. Its warhead carried 3,400 pounds of high explosives. It was to have been launched for take-off on a 400 foot ramp, whereupon a guidance system would guide it to the enemy target. The war ended before any of these were placed into quantity production and only a few models were built

Experimental Glider for Buzz Bomb Program. This tiny glider was built to test the efficiency of the flying wing design in jet bombs. Elimination of the body and tail on this midget resulted in a clear glider which demonstrated the high efficiency for which the Northrop Flying Wings were noted. After being towed aloft, it was even capable of slow rolls and free flight. The twin jet propelled Buzz Bomb was built to the same design as the MX-334, except for the elimination of the pilot's seat and canopy. This glider is shown here as used for instructional purposes at Northrop Institute, Los Angeles, California.

Northrop MX-543 "Flying Bat"

This twin jet buzz bomb was powered by two General Electric turbojet units. It was the first venture by the Northrop Corporation in the design and manufacture of robot bombs for the Army. A number of these jet bombs were supplied to the Army before Northrop switched to the JB-10 which utilized a German-type pulse jet engine. Two 1,000 pound bombs were carried within the bulges on each side of the cockpit area. A cockpit was mounted on this test vehicle to gather research data.

Northrop JB-1 "Jet Bomb"

Northrop's JB-1 was a remote controlled jet bomb. Ground launched with preset controls, it carried a payload of two, 2,000 pound bombs over a range of 200 miles. It was powered by two turbojet engines. Take-off was accomplished from a 400 foot ramp. Five 10,500 pound thrust rockets were on the launch ramp to propell the JB-1 in its initial take-off. This model bore the Northrop designation of MX-543.

Northrop JB-10 "Buzz Bomb"

Northrop's JB-10 "Buzz Bomb" carried nearly two tons of high explosives. A copy of the German V-1 pulse jet engine powered the JB-10. The inner portion of the flying bomb was built of cast magnesium and the outer wing panels were of aluminum construction.

XP-79

In 1942 work had begun in the United States on the development of a rocket-driven interceptor of even more advanced flying wing design. In this design the pilot lay prone. This work had stemmed from a proposal made to the Air Force Air Material Command in September, 1942 by John K. Northrop.

When John Northrop made his proposal to the Air Force, rocket motor development in the United States was very much in its infancy. Service interest had been largely confined to the application of rockets as take-off-assistance devices. The AFAMC had begun work on JATO (Jet-Assisted Take-Off) units in 1939.

Aerojet had already begun work on acid-aniline rocket motors which were capable of substantially longer burning times than those required for JATO tasks. They were also capable of being re-ignited an almost unlimited number of times in the air, and they offered regulated thrust. John Northrop believed that a power plant capable of giving 2,000 pounds of thrust would be ideally suited to the airplane he had in mind -- an airplane in which the physical dimensions of the human frame was the limiting size factor. For this reason a prone position was desirable for the pilot. Such pilot accommodations allowed the cockpit to be no more than a slight thickening of the wing section; it had the advantage of showing the minimum silhouette to enemy gunners; enabled the pilot to withstand twelve "g" acceleration and reduced the strain imposed by violent maneuvers and sudden pull-outs.

Northrop proposed that the primary structural material should be heavy welded magnesium plate. He appreciated the fact that red fuming nitric acid, used by the Aerojet rocket motor as an oxidant, had a highly corrosive effect on magnesium, but he considered the protection of the integral fuel tanks from battle damage the primary importance. The prone pilot position and the use of magnesium in the structure of the aircraft were radical features. But no feature was more radical than the flying wing configuration.

Apart from the thick magnesium alloy skins, armor was to be provided in the form of a quarter-inch face-hardened steel plate, inserted just inside the wing leading edge at a forty-five degrees to the chord plane. It was to extend from the side of the aniline tank adjacent to the cockpit to the outboard side of the main nitric acid tanks. Also included was a bullet-proof glass to protect the pilot.

It became apparent that Aerojet was having difficulty in finalizing the design of the "Rotojet" rocket motor, and Avion advised the Air Material Command that delivery of the first XP-79 would be postponed from the contractual delivery date of 30 September until December 15, 1943. Subsequently, in September, the estimated delivery date was again revised -- this time to January 15, 1944.

Delay followed delay. Avion submitted numerous generalized reports on the XP-79 and various aspects of its development to the Air Material Command, who considered them to be of no more than academic interest insofar as the project was concerned. They finally instructed Avion to terminate "all unnecessary work" in order to concentrate its effort on building a flyable airplane.

In January, 1943, procurement of three prototype airlanes under the designation XP-79 was initiated. At that stage of the war, Northrop Aircraft was severely restricted in its choice of sub-contractors. But they selected Avion incorporated, which was occupying the Gaffers and Sattler Stove Factory in Los Angeles. Its staff included a number of engineers formerly with Vultee Aircraft. At the same time, the decision was made to build three full-scale flying models of this interceptor.

The detailed model specification for the XP-79, submitted to the Air Force Air Materiel Command on April 1, 1943, envisaged an airplane with a loaded weight of 11,400 pounds; a wing spanning thirty-six feet respectively; an aspects ratio of 5.07; and a gross wing area of 255 square feet. Welded magnesium alloy was to be used throughout the basic wing structure, which was a monocoque structure with wing skin varying in thickness from three-quarters of an inch at the leading edge to three-sixteenths at the trailing edge. The center section, eight feet in width, was to house the pilot's cockpit, the rocket motor, and the fuel tanks and armament. The major portion of the outer panels was to be occupied by the oxidant fuel tanks.

All movable control surfaces were made of welded magnesium alloy. These were to consist of elevons for pitch and roll extending over some sixty per cent of the wing and split-type maneuver brakes extending outboard from the center-line of the airplane over approximately forty per cent of the semispan. The pilot was to be provided with a crossbar with hand grips for operating the crossbar on its horizontal axis. The maneuver brakes were to be controlled manually by foot pedals with sufficient power boost to reduce the operating loads to a reasonable level. The boost was to be obtained from dynamic air pressure acting on an internal bellows within the maneuver brakes themselves.

The pilot's cockpit was to be sealed to form a pressurized capsule capable of withstanding 2.75 pounds per square inch, but no means of actually pressurizing the cockpit were to be provided. Armament was to comprise four 0.5 inch calibre M2 Browning machine guns -- two on either side of the longitudinal axis of the airplane just outboard of the fuel tanks.

XP-79B

In 1944, Northrop was awarded a contract to build the XP-79B "Flying Ram" fighter. This design also featured a prone positioned pilot in the cockpit. It was hoped that this design would reduce the strain on the pilot during violent maneuvers and pull outs and would also present minimum silhouette to enemy gunners.

With the availability of jet engines, a decision was made to power the third model with two Westinghouse 19-B axial flow turbojets. The XP-79B arrived at Muroc Army Air Base in June, 1945. Test pilot for this program was Harry Crosby. This unique aircraft had a very low profile. Sitting low to the ground, it had four retractable wheels. It also had twin vertical fins atop the jet engine exhausts. It was of welded magnesium construction, and, as a Flying Wing, featured the air bellows operated split flap wing tip rudders outboard of the elevators, like the XP-56. The design construction of the wing was heavily reinforced so that it could be used to ram enemy bombers and slice off their tails. The first test flight was made September 12, 1945.

During its first test flight, the XP-79B made a successful take off from Muroc Army Air Field and was seen climbing for altitude. Shortly after this, however, it crashed to the ground, killing test pilot Harry Crosby. No further aircraft were ordered, and the project was cancelled.

The XP-79B had a wing span of 38 feet, a length of 14 feet, and an overall height of 7 feet. It had a gross weight of 8,670 pounds and carried a fuel load of 300 gallons. The XP-79B's top speed was 500 miles per hour and its range was 990 miles. It carried two Westinghouse 19-B engines buried within the wing and delivered 1,345 pounds of thrust.

Only one XP-79B "Flying Ram" was built. It was the first airplane intended to down enemy bombers by slicing off their tail assembly with the strong leading edge plate.

XB-35 Flying Wing

The XB-35 was the first in a series of large Flying Wing Bombers. It was a bombardment type aircraft of exceptionally long range and heavy load capacity. Two XB-35 and 13 YB-35's were initially scheduled to be built under Army contracts. But only six of the big wing bombers were completed and test flown. The XB-35 had a wing span of 172 feet and a wing area of 4,000 square feet. It was capable of operating under overloading conditions at a gross weight of 209,000 pounds or 104.5 tons.

Prior to the XB-35 design, Northrop had flown twelve tailless aircraft designs since the company was founded. Northrop engineers learned much about these experimental models and had made hundreds of additional models for wing tunnel tests. For the first time in the history of military aeronautics, the flying wing contributed many advancements to military design in the bombardment category. Five of the leading advantages were:

1. The low drag, high lift feature of the Flying Wing meant that in practice the XB-35 could transport any weight faster, farther and cheaper than an aircraft of conventional design.

2. The simplicity of construction of the XB-35 Flying Wing presented few structural complications. It cost less to build since it was built as a single unit in which the structure extended through from tip to tip with no added tail or fuselage to build.

3. It had better weight distribution. Compartments along the span could distribute the bomb load and weight more evenly over the wing surface which supported it. This also would have applied to cargo if a cargo version had been built. It eliminates the need for excessive structural weight such as is necessary in conventional bombers where the weight is concentrated in the fuselage and must be distributed over the wing from this point through the use of heavy structural members.

4. For ease in loading and unloading, cargo could be placed in span wise compartments where any part of the plane's load was easily accessible through its own cargo or bomb bay doors. This provided direct access to all portions of the cargo.

5. The XB-35 for military purposes presented a smaller target while engaged in either offensive or defensive operations. Comparisons between the Flying Wing and conventional aircraft is also illustrated by performance figures obtained from Northrop and backed by many years of research tests in which planes identical in scale to the XB-35 were used. The XB-35 was built to be 20 percent faster than conventional bomber aircraft with identical loads and horsepower. Normal crew for the XB-35 was nine men -- a pilot, co-pilot, bombardier, navigator, engineer, radio operator, and three gunners. Cabin space was available for six more crew members, who could substitute on long-range missions. Folding bunks were built in the XB-35 to accommodate the off-duty crewmen. These fifteen men were housed entirely within the wing foil itself. The XB-35 was built of a new aluminum developed by the Alcoa Company. Tests showed this material to be considerably stronger than previous metals used. Gasoline was carried in bullet proof, leak proof fuel cells within the wing of the XB-35 and additional range was built in by the addition of additional fuel tanks in the bomb bay and other wing compartment areas. The wing section of the XB-35 was 37-1/2 feet long at the center, tapering to slightly more than nine feet at the wing tips. It swept back from center to tips making the overall length of the ship slightly more than 53 feet. The XB-35 stood over 20 feet tall when at rest on its tricycle landing gear. It was equipped with 5'6" dual wheels on the main gear and a 4'8" wheel on the nose gear.

The XB-35 received its initial start from the Army Air Corps in September, 1941, following a visit by Assistant Secretary of War Robert Leavitt, General H. H. Arnold, and Major General Oliver P. Echols. At this time, Northrop submitted a preliminary design for the XB-35 to Wright Field. Army officials were more than satisfied because they awarded a contract for the altitude performance with a pressurized cabin for the crew and a more than adequate bomb load. There would not have been an XB-35 Flying Wing Bomber had it not been for the enthusiasm and foresightedness of men like John Northrop, General Arnold, General Echols, and Mr. Leavitt.

In November, 1941, two XB-35 prototypes were ordered by the Army Air Force. Preliminary design work began on the XB-35 early in 1942, and on July 5th of that year the mock-up Board from Wright Field inspected a full size wood mock-up of the center section and a portion of the left wing. Official approval was given. Northrop built a new bomber plant at the end of the company's grounds in Hawthorne, California. This plant was completed in January, 1943 and fabrication of the parts for the first XB-35 were begun. The complexities of design in building the B-35 represented an operation of major size. The Martin Company played a very important role by providing wind tunnel and other research data for the XB-35 design. To provide flight test data, Northrop built four, sixty foot wing span N9M Flying Wings. While the XB-35 was taking shape in the Hawthorne factory, thousands of test flights were run on the N9Ms to gather data which could be worked into the XB-35 design. This time-saving device enabled Northrop engineers to cut development time from the XB-35. Over the next three years flight tests were conducted at Muroc Army Air Base to gather data while the XB-35 was taking shape.

The XB-35 was originally planned to have three-bladed Hamilton Standard propellers; however, during testing, it was found that the four-bladed propeller was more efficient. Note that the first engine in the foreground has the three-bladed propeller still attached while the other three engines have been modified to carry the Hamilton Standard four-bladed propeller. This open house view for Northrop employees was the first public showing of the flying wing design.

A XB-35 is shown in flight over Muroc Dry Lake. Eight four-bladed counter rotation propellers drove the Flying Wing through the air. This view clearly shows the turret blisters on the upper portion of the wing envelope.

The XB-35 used elevons which were installed on the trailing edges with landing flaps, trim flaps and rudders. The elevon was a Northrop invention in which the function of both elevators and ailerons were combined. Further control was achieved by wing slots. The XB-35 was first flight tested during the summer of 1946 by Northrop's Chief Pilot, Max Stanley. One YB-35 was slated to be turned over the the Navy and was to be designated XB2T-1; however, this did not take place. The Navy was to have flown and tested the aircraft for research purposes, but this project was cancelled.

After the prototype XB-35 models were built at the Hawthorne plant, a war time plan was begun to build 200 B-35 Flying Wing bombers by Martin Aircraft Corporation at Omaha, Nebraska. Martin had previously built B-29 bombers, and it was expected that these would be phased out in favor of the B-35 Flying Wing Bombers. This program was dropped, however, with the conclusion of WorldWar II. The first XB 35 flew on June 25, 1946. The power plants of the XB-35 were Pratt-Whitney "Wasp Major" engines. It had two B-4360-17 and two R-4360-21 engines of 3,000 horsepower each with double turbo super chargers and eight coaxial counter rotating four bladed Hamilton Standard pusher propellers. Tests were also run on three-bladed propellers, but the four-bladed models proved to be the most efficient. The aircraft was designed so that other types of engines could be adapted as they were made available.

The XB-35 was to have carried twenty 50 caliber machine guns, but these were not installed on the number one prototype XB-35. Seven remote-control gun turrets were aimed from central firing sighting stations behind the pilot and on top of the cone protruding from the trailing edge. Four gun turrets were fitted above and below the center section. Two gun turrets were visible outboard of the engines -- one on top and one below. Four guns were to have been placed in the tail cone. The first two XB-35 and sole YB-35 Flying Wings had "Wasp Major" engines. But on June 1, 1945, orders were issued to have the next two finished with Allison J-35-A-5 jet engines. These models were known as YB-49 jet Flying Wing Bombers. On June 25, 1945, American Aviation took a bold step into the future. On this date the world's first all Flying Wing Bomber took to the air from Northrop Field at Hawthorne, California. The crew for this first flight were pilot, Max Stanley and Dale Schroeder, flight engineer.

This is a rearward view of the number one prototype XB-35 at Muroc Army Air Base. Note the crew compartment entrance hatch in the open position in the bottom of the wing.

Following the flight of the first XB-35, (S/N 42-13603), a second prototype XB-35 first flew in November of 1946. The second prototype XB-35, (S/N 42-38323) soon joined her sister ship on the ramp at Muroc Army Air Base. When the XB-35 was completed and had made its maiden flight, further tests were conducted to prove the feasibility of the design. The two Pratt-Whitney R-4360-21 Wasp Major engines drove the inboard contro-rotating propellers and the two R-4360-17's turned the outer pair of propellers. Each engine developed 3,000 horsepower. The metal housing over the propeller shafts offered enough area to counter-act any yawing tendency of the Flying Wing design, and counter rotating propellers gave additional stability by eliminating torque problems. Cooling air for the radial engines were ducted from openings in the leading edge of the wing. The XB-35 had a maximum speed of 395 miles per hour and a cruising speed of 183 miles per hour.

The XB-35 Flying Wing Bomber is shown on the ramp at Muroc Air Force Base. Note the six gun turret blisters on the wing; there are three on top and three on the lower surface. The pilots were located in the bubble canopy situated on top. The bombardier was positioned in the center of the nose.

This is a rare photo of nine Northrop Flying Wing Bombers. Many people do not realize that more than one or two prototypes were built of this design. Here, for the first time, is actual proof of their existence. Two of the big wing bombers are undergoing modifications from XB-35's to Flying Wing B-49 Jet Bombers. The entire project was later cancelled by the Air Force.

INSIDE THE FLYING WING

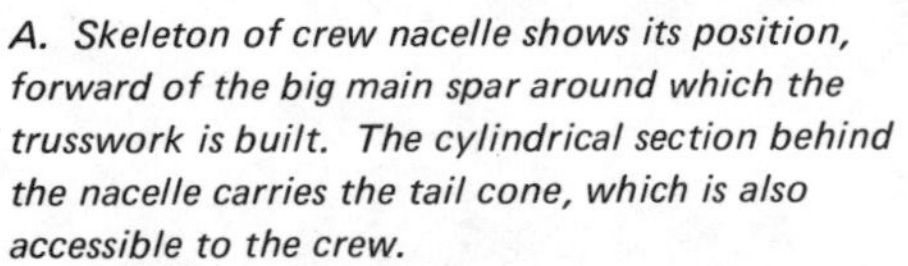

A. Skeleton of crew nacelle shows its position, forward of the big main spar around which the trusswork is built. The cylindrical section behind the nacelle carries the tail cone, which is also accessible to the crew.

B. Instruments and controls of the plane are shown in a full-size wooden mockup. Pilot's wheel is at left, and his head projects up into Plexiglas blister directly above it. Co-pilot's wheel, at lower right, is behind windows in leading edge of wing.

C. Structural strength of the giant wing is indicated by framework taking shape around the center nacelle. This stage of construction also illustrates relative simplicity of building a plane in which the fuselage is an integral part of the wing.

A.

B.

C.

D.

E.

F.

D. Taken inside the plane itself, the photo shows flight engineer's station and the mass of instruments he has to observe to check on plane's performance. The engineer rides backward, about 10 feet behind the co-pilot's seat. The navigator sits near him at the table in left foreground; table is also shown at bottom right of picture.

E. This view, looking forward to the co-pilot's station, shows location of bombardier's station. The square opening just below and to the right of the control wheel is for a glass panel in the forward wing surface, behind which the bombardier sits or kneels to operate bombsight.

F. This view is the pilot's station, as seen from center of the crew nacelle. The overhead blister can be seen through the diagonal braces, which partly obscure the control wheel and instrument panel. Table at bottom left is the radio operator's station, shown with some of the flight-test equipment required for early trials of the Flying Wing.

The third Northrop prototype Flying Wing bomber was designated YB-35, (S/N 42-102366). More were on order for further testing, but trouble developed in the engine reduction gear box arrangement and in the propeller governors of the first three prototypes. Further development of the B-35 project was abandoned in favor of the YB-49 all jet Flying Wing bomber. Of the original thirteen YB-35's ordered, four were scheduled to be used to supply spare parts for the extensive flight test program planned. A decision to convert the remaining B-35's to more modern propulsion was based on the outstanding performance recorded by the YB-49 jet bomber.

Turbodyne

One YB-35 Flying Wing was to have been used as a test bed for the powerful Northrop XT-37 "Turbodyne" engine developed by the Turbodyne Corporation, a subsidiary of the Northrop Corporation. This project was dropped, however, when the Air Force ordered Northrop to divest itself from the manufacture of engines. All patents and manufacturing rights were later sold to the General Electric Company.

YB-49 and YRB-49A

With the introduction of the YB-49 to flight status in the fall of 1947, it was anticipated that this model would prove to be the most successful flying wing aircraft. A contract ordering the conversion of two YB-35 Flying Wings was issued on June 1, 1945. This specification called for the installation of eight Allison J-35-A-5 turbo jet engines of 4,000 pounds thrust each.

The number two prototype YB-49 Flying Wing is shown in flight over Muroc Dry Lake with the silver wing glinting in the early morning sun. The mammoth YB-49 flies over Muroc Dry Lake on one of its many research missions. At the time it was the world's longest and largest ranging jet airplane. The YB-49 demonstrated numerous advantages in the flying wing design. It was capable of traveling farther and faster and of carrying heavier loads than conventional airplanes of comparable size.

The YB-49 Flying Wing weighted 88,100 pounds when it was rolled out of the factory. The normal loaded weight was 205,000 pounds. This could be boosted to 213,000 pounds if necessary. The first flight was made on the number one prototype YB-49 on October 1, 1947. This aircraft, S/N 42-102367, was sucessfully flown from Hawthorne, California to Muroc Air Force Base by Northrop's cheif test pilot, Max Stanley. Subsequent test flights pushed the plane to a top speed of 520 miles per hour and placed the ceiling surface at 42,000 feet. It held fuel capacity of 17,545 gallons and had a range of 4,450 miles. The YB-49 Flying Wing jet bomber was capable of carrying over 36,760 pounds of bombs for a distance of 1,150 miles. A normal 10,000 pound load could be carried for an estimated 4,000 miles on 6,700 gallons of fuel. This was less than half the range of the B-35. This was due to the great quantity of fuel required by the turbo jet engines; however, a 100 mile increase in speed was gained in the YB-49.

Over twenty months of research and flight testing was conducted on the two YB-49 prototypes at Muroc Air Field. During this time, payload test and endurance records were broken. Contracts were placed for thirty RB-49's and for converting the remaining ten B-35 airframes to YRB-49A strategic reconnaissance type aircraft. During two years of flight testing, stability problems encountered in the YB-49 design had not been overcome and further tests were planned. A new autopilot was to be designed and built into the aircraft. Before this could be done, the second prototype crashed:

An unusual view from below the YB-49 shows four vertical rudders plus the central tail cone. The clean aero dynamic lines of the wing are apparent as the YB-49 flies majestically above the Mojave Desert of Southern California. At this time it was heralded as the airplane of the future and was perhaps the most unusual appearing of all post-war designs.

On the morning of June 5, 1948, Captain Glenn Edwards, a test pilot at Muroc Air Force Base, was co-pilot on the number two prototype YB-49. This aircraft had several aft C.G. stability test which had to be completed before this aircraft was accepted by the U.S. Air Force. Scheduled stability tests were conducted this day at 40,000 feet at a location just north of Muroc Dry Lake. Captain Edwards radioed to base after his tests were completed; they were dropping down to 15,000 feet. The pilot was not heard from again. Witnesses said they saw it tumbling down out of control just north of Highway 58 and crash. Captain Edwards and the entire crew were killed. What exactly happened is not known. The clean all wing design of the YB-49 enabled the aircraft to reach very high speeds. In descending from high altitudes, the YB-49 could easily surpass "red line" (not to exceed) air speed. It is assumed that Captain Edwards was attempting to investigate the stall characteristics of the aircraft, and the "not to exceed" limits of the aircraft were exceeded while descending from 40,000 feet. In exceeding these limits, the outer wing panels were shed and the aircraft tumbled in three pieces to the ground.

Getting airborne, the number one prototype YB-49 lifts off the runway. Note the tricycle landing gear has not been retracted. Company personnel and neighbors line the field to watch this giant take wing.

The number one prototype YB-49 Flying Wing jet bomber prepares for its initial takeoff from Hawthorne Air Field to Muroc Air Force Base on October 21, 1947.

Up! Up! and away! – The Northrop YB-49 takes off from Hawthorne Air Field. Eight Allison J-35 turbojet engines gave a combined thrust of 32,000 pounds of thrust to power this giant. Max Stanley, YB-49 test pilot was concerned at first about the short 5,000 feet of runway at Hawthorne. The plane took off easily with room to spare.

Here is the number one prototype YB-49. Note the four small rudders with wing fence extensions which gave this design more directional stability. The YB-49 had a very sharp turning radius and could literally turn inside some of the modern jet fighters of its day. The YB-49 had a top speed of 520 miles per hour.

The YB-49 crash site can still be seen today. Located ten miles east of the city of Mojave and just north of Highway 58, the scorched earth and a few bits of plastic and metal are all that remain as a monument to this amazing design. This crash rang the death knell for the YB-49 Flying Wing program. Muroc Air Base was later named in honor of Captain Glenn Edwards. Today it is known as Edwards Air Force Base, Flight Test Center of the United States Air Force. Only two YB-49's were completed and test flown. It had a load factor of two and a top speed 520 miles per hour and a service ceiling of 42,000 feet. The YB-49 had eight Allison J-35-A-5 turbo jet engines. These were located in banks of four at either side of the wing, and the only protuberances were the exhaust along the trailing edge. Air intakes were mounted in the leading edge of the wing. All gun turrets, except for the tail cone guns, were eliminated. Four verticals stabilizing fins were fitted. The regular crew of seven were housed entirely within the seven foot thick wing center section. The pilot was located in a plastic bubble canopy near the leading edge for excellent visibility. For long flights provision was made for an off-duty crew of six members who had quarters in the tail cone just aft of the flight section. One YB-49 averaged 511 miles per hour during a 2,258 mile flight from Muroc, California to Washington, D. C. Flights of over nine hours duration had been recorded. These performances give support to the flying wing design. At its time it was one of the world's longest ranging jet aircraft.

Flying scenes of the YB-49 aircraft were featured in a Hollywood film based upon H. G. Wells' "War of the Worlds". If one is privileged in viewing this on a late show on television, you will see an extraordinary art form of flying, banking, rolling and turning as graceful as any plane that flew. Watching the Flying Wing fly and perform in color is an experience not easily forgotten. This author had the privilege of viewing the YB-49 flying at March Field near Riverside, California in May, 1947. Its grace and beauty in the air will always be a sight to be remembered, but what impressed this writer was its ability to turn inside a top Air Force Fighter of the day -- the P-80 "Shooting Star".

The first prototype YB-49, S/N 42-102367, was burned and destroyed during a maximum load, high speed, taxiing accident at Edwards Air Force Base. This left the YRB-49A, a strategic reconnaissance aircraft as the sole remaining Flying Wing. This photo reconnaissance aircraft had four 5,000 pound thrust Allison J-35-A-19 engines in the wing and two more suspended in pods below the wing. Photographic equipment was installed in the tail cone bay just below the center section. It had an empty weight of 88,500 pounds and a gross weight of 117,500 pounds and a maximum gross weight of 206,000 pounds. It had a top speed of better that 550 miles per hour and it was the same dimensions as the YB-49.

The clean lines of this flying wing was broken by two engine pods suspended below the wing leading edge to carry two Allison J-35 turbo jet engines. The YRB-49-A first took to the air on May 4, 1950. Tests were conducted at Edwards Air Force Base for a period of time, but stability tests in the program hampered further production. Plans were made to install a stabilizing device made by Minneapolis Honeywell to overcome stability problems. This was the same type of device installed in the B-47 Stratojet Bomber for stability guidance. The YRB-49A was designed as a fully operational photo reconnaissance plane. It had useful loads and better performance speed and range than the YB-49. In early 1952, the huge flying wing was flown to Ontario International Airport to the Northrop facility where the installation of this stability device was to be installed. Funding for this project was dropped, however, by the Air Force, and the aircraft remained in dead storage for a period of time. In October, 1953, the Air Force ordered the YRB-49A to be scrapped. Crews were sent from nearby Norton Air Force Base to complete the salvage project. The aircraft was towed from the Northrop facility at the southeast end of the field to a site adjacent to the Air National Guard area where it was dismantled and cut up for scrap. The huge ship had been sitting in the grape vineyards weathering the onslaught of dust, wind, rain and sun for some time. One major Los Angeles newspaper stated: "It is hard to believe we have scrapped a design which is as modern as the planes today. Perhaps we will have to wait twenty years or longer before we find out the true significance of our mistake.'' True, the wing did have its share of problems; however, with adequate funding, these problems could have been worked out to prove that the wing did hold a place with the Air Force. The Flying Wing was years ahead of its time, but it was caught in a budget squeeze between the faster B-47 Stratojets and the long-range B-36 inter-continental bombers. Politics also played and important part in the defeat of the B-49 Flying Wing. Perhaps, some day, John Northrop will release the full story.

The YRB-49A had a top speed of 512 miles per hour at 40,000 feet. It had a range of 3,500 miles non stop without aerial refueling.

A comment must be made on the B-49 design competition: During this period, 1948 - 1952, a former Northrop competitor employee became Secretary of Defense under President Truman's administration. Being a former officer of a competing firm to Northrop, some significance in the Air Force's selection of the B-36 over the B-49 should be noted. One day a high government official was sent from Washington to visit John Northrop. A meeting took place whereby John Northrop was told that his B-49 Flying Wing Bomber would be produced for the Air Force; however, it had to be built by Convair in Fort Worth, Texas or the Air Force would not buy his Flying Wing design. John Northrop replied that he and his employees had built the Northrop Company and organization over the years and he owed a debt of loyalty to his employees. He stated that if the aircraft was to be built at all it would be built by his people at the Northrop factory in Hawthorne, California. The rest is history. The Air Force ordered the B-36. The significance of this fact should be made clear to future generations investigating the story of the Northrop Flying Wing.

A YRB-49A and T-33 chase plane is shown over the Mojave Desert during a test flight. Note the position of the elevons in this right-turn maneuver.

This is a YB-49 side view showing air inlet for eight J-35 turbojet engines, wing slots, and pilot tube. The YB-49 had been placed in Air Force service, it would have had a longer service career than any previous bomber because it could have been easily modified as more powerful turbojets were developed.

The first flight of the YRB-49A took place on May 4, 1950 from Hawthorne, California. It is shown here taxiing out to the ramp for take-off.

The YRB-49A was a reconnaissance version of the Flying Wing Bomber. Housed in its tail cone were photographic cameras. It carried six Allison turbojet engines. Four were mounted in the rear trailing edge of the wing, and two were mounted in pods under the leading edge of the wing.

This is a side view of the YRB-49A. Two J-35 Allison engines were mounted in wing pods below the leading edge of the wing. Overall dimensions were the same as the YB-49. This model was intended for use as high speed photographic aircraft for the Air Force.

Here is an unusual view of the YRB-49A in flight. The externally mounted twin turbojet units below the wing gave this model an unusual appearance. Four more Allison turbojet units are mounted at the turning edge of the wing — two within each side of the twin rudders.

A YRB-49A Flying Wing takes off from Edwards Air Force Base. This unusual view shows thrust being generated by the six Allison J-35 turbojet engines upon lifting off the ground. The YRB-49A was the last of the Flying Wing Bombers to be built. This aircraft came to a sad end. It was scrapped at Ontario International Airport in October, 1953. This act heralded the end of the Northrop Flying Wing in this country.

The Northrop YRB-49A, serial number 42-102376, was originally the eighth YB-35 Flying Wing Bomber. A contract was issued by the Air Force to Northrop to modify this model to the YRB-49A configuration. It flew for over three years at Edwards Air Force Base.

The YRB-49A is shown above Mojave Desert on a test flight. Stability problems were overcome by the installation of a Honeywell automatic pilot.

This is a close-up view of the Allison J-35 turbojet of 5,000 pounds thrust as shown on the YRB-49A. These engines are now held by the Aeronautical Shop at Chaffey College, California.

This large all metal Flying Wing Jet reconnaissance aircraft was a remarkable sight over Southern California during the early 1950's. It was the first Northrop Wing to be equipped for full tactucal operations. It carried a crew of six men: pilot, co-pilot, flight engineer, photo navigator, radar navigator and photo technician.

FLIGHT CONTROL OPERATIONS

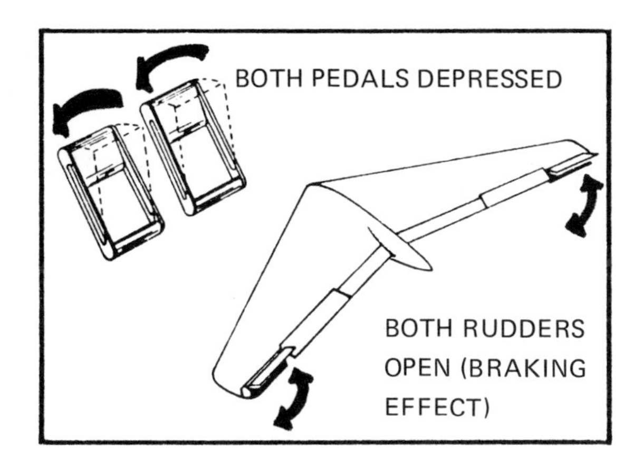

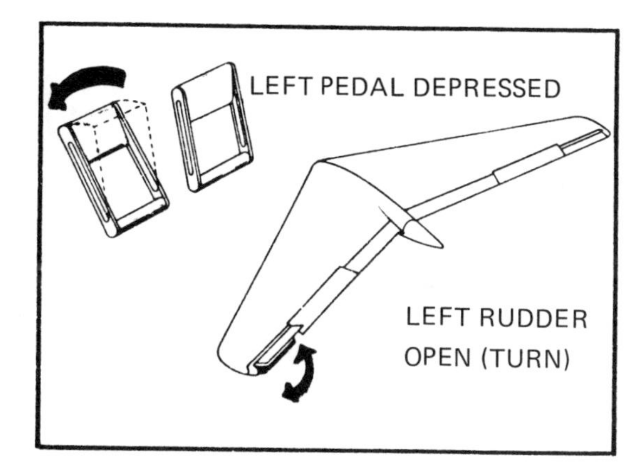

The flight controls in the cockpit operate in a normal fashion, and differ only in minor degree from their conventional counterparts. Foot pedals operate the rudders which consist of double split flaps (something like dive flaps) located at the wing tips. When a rudder pedal is depressed the flaps open to produce drag at the required wing tip. Both pedals may be pushed to open both rudders to increase the gliding angle or reduce the airspeed. This airbrake feature is the only variation from a conventional cockpit arrangement.

The rudders form a portion of the trim flaps, which are located at the wing tips, and are adjusted up or down to trim the airplane longitudinally in the same manner as one would use an elevator trim tab, or an adjustable stabilizer.

Elevons, combining the function of elevator and aileron are located along the trailing edge on each wing inboard of the trim flaps.

When deflected together in the same direction by moving the control column fore and aft, the elevons cause the airplane to descend or climb exactly as normal elevators operate.

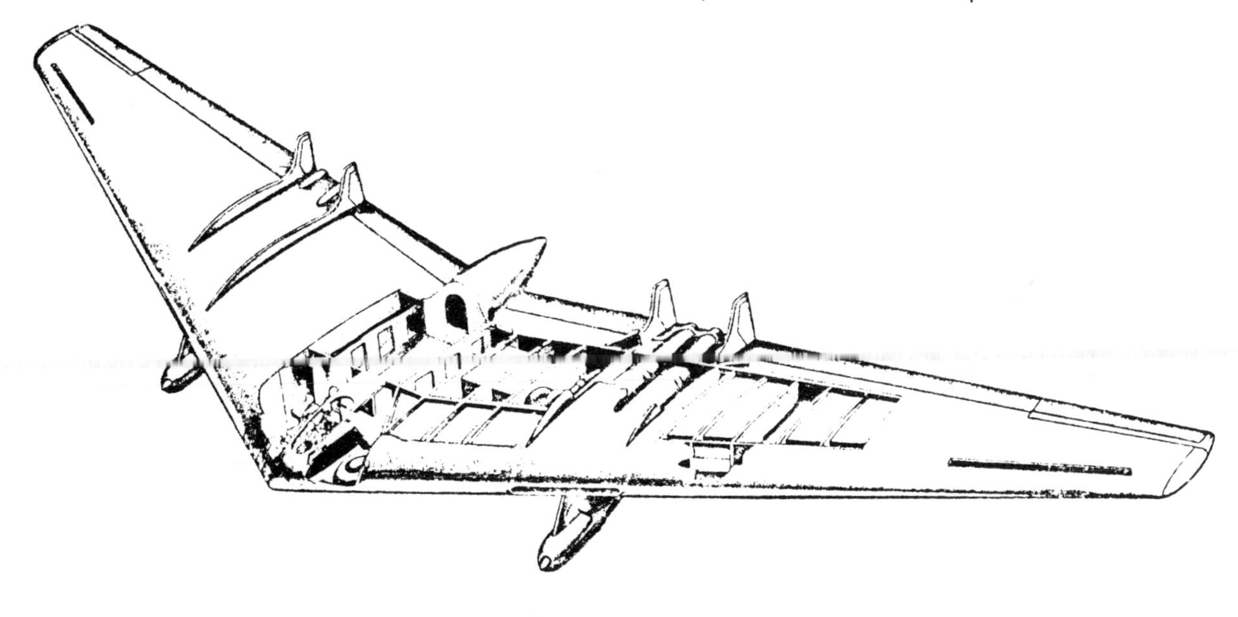

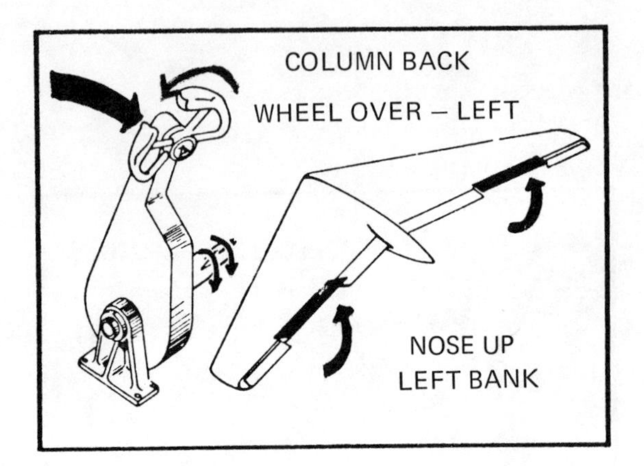

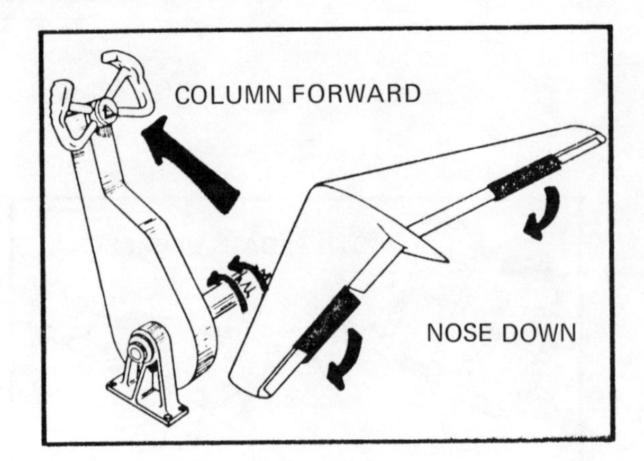

When moved in opposite directions by rotating the control wheel to right or left, they cause the airplane to bank in a fashion identical to that caused by conventional ailerons. The elevons are connected to the control column in such a manner that completely conventional pitch and roll control is obtained by moving the column, or rotating the wheel.

Landing flaps are located in conventional position at the trailing edge near the center of the airplane and serve to increase the lift (for slower landing) increase the drag (for steeper approach) and decrease the angle of attack required to achieve a proper lift coefficient for landing. When depressed, they cause a nose-down effect which is trimmed out by an upward setting of the trim flaps at the wing tips.

The size of a big bomber often requires that all control surfaces have hydraulic or electrical boost systems to supplement the pilot's strength. On the B-35 and B-49 airplanes there are complete dual hydraulic systems, in addition to an electrical trim system which is sufficient for control in emergency flight.

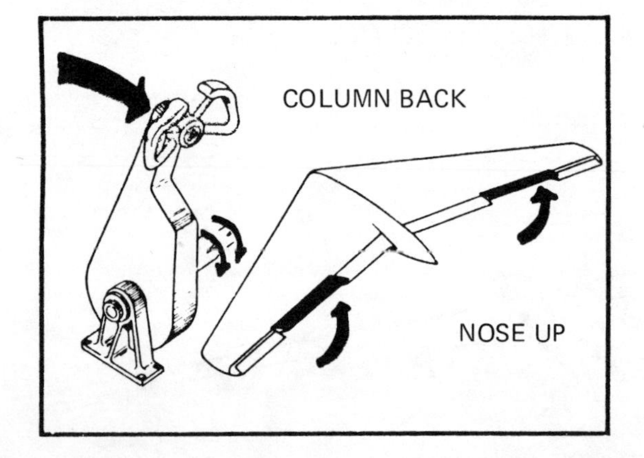

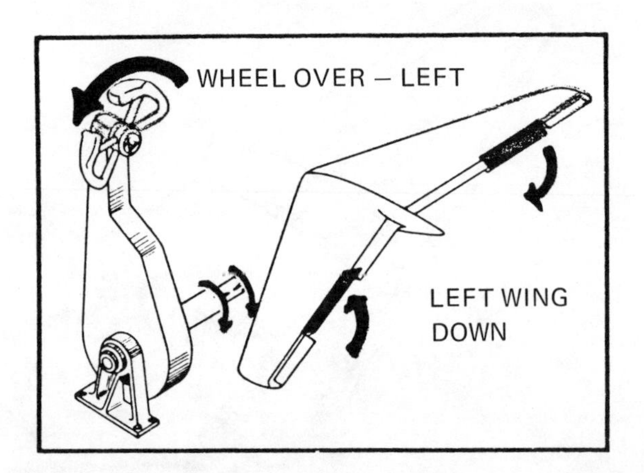

Northrop X-4

The last Northrop designed flying wing to be built before John Northrop retired from active design work was the X-4. The X-4 was a miniature flying wing laboratory intended to explore the stability of aircraft of swept back design and all wing configuration at sonic speed ranges. The wing surfaces of this aircraft were swept back, and the tail surfaces consisted of only one vertical fin and rudder. It had no horizontal stabilizer. The test pilot of the X-4 was Charles Tucker. The X-4 was 23 feet, 4 inches in length and it had a wing span of approximately 26 feet, 4 inches. It was 15 feet high and had a design gross weight of approximately 7,000 pounds. Extensive flight programs were carried out on the X-4 by the Air Force and NASA at Edwards Air Force Base. Only two models of the X-4 were manufactured by Northrop at its Hawthorne factory. Today these aircraft still exist. One is on display in the Cadet Courtyard at the Air Force Academy in Colorado Springs, Colorado. The other is at the Air Force Museum in Dayton, Ohio. The X-4 had twin jet engines. It carried two Westinghouse J-30 Turbo jet axial flow jet engines, each with 1600 pound thrust. It had a top speed of 550 miles per hour.

Northrop's X-4 (S/N 6677) of 1948 was one of two X-4's built for the Air Force. It was a miniature flying laboratory intended to explore the stability and flight of aircraft with swept back and all wing configuration at sonic speed ranges.

The X-4 was approximately 20 feet in length and had a wing span of approximately 25 feet. It was 15 feet high and had a gross weight of 7,000 pounds. Extensive flight programs were carried on with the X-4 at Edwards AFB and later with NASA. One of these aircraft is on exhibit at the Air Force Academy at Colorado Springs and is a cadet mascot. The other is on exhibit at the Air Force Museum.

In 1947 the Northrop Corporation received a contract from the U.S. Air Force to build two experimental research aircraft designated X-4. These aircraft were completed and test flown in 1948. These aircraft were miniature flying laboratories intended to explore stability and flight of aircraft of the swept back and all wing configuration at the sonic speed range. The wing surfaces were swept back, and tail surfaces consisted of only vertical fin and rudder. It had no horizontal stabilizer.

MAX STANLEY ferried bombers and transports to British West Africa during the early part of WWII. He test flew the XB-35, the YB-35, the YB-49 and the YRB-49A.

JOHN MEYERS gave up law practice to ferry bombers during the early part of WWII. He made his first flights in the N9M Flying Wing, the NP-56 and the P-61.

MOYE STEPHENS, as secretary of the Northrop Company, flew early N1M Flying Wing tests and N9M series tests. He flew Richard Hallibarton around the world in 1931-32.

HARRY CROSBY, noted early National Air Race pilot at Cleveland, was an airlines pilot in Central America and a builder of the world's smallest all metal plane. He test flew the XP-56, the N9M, the Northrop MX-324 and the XP-79B "Flying Ram".

L.A. "SLIM" PERRETT was a barnstormer and motion picture pilot. He later flew for airlines and was a former AAF instructor pilot. He test flew the P-61 and the XF-15 Reporter.

DICK RAMULDI was also a barnstormer and motion picture pilot. He flew for the airlines. He taught Captain G. Allen Hancock and film star Wallace Berry to fly.

SPECIFICATIONS

YEAR	MODEL	ENGINE	POWER	WING SPAN	LENGTH	WING AREA SQUARE FEET	ASPECT RATIO	REG. SER. NO./ NO.	MAX. SPEED
1929	Twin Boom	(1) Menasco A-4 (2) Cirrus	90 H.P. 100 H.P.	30'6"	20'	184	5:1	X-216H	100 m.p.h.
1940	N1M	Lycoming Franklin	(2) 65 H.P. (2) 120 H.P.	38'	20'	300	4.75:1	NX-28311	200 m.p.h.
1942	N9M	Menasco (CGS-4)	(2) 275 H.P.	60'	17.79'	490			
1943	N9MB	Franklin (0-540-7)	(2) 375 H.P.	60'	17.79'	490		.0039	
1943	XP-56 (N2B)	P&W R-2800-29	2000 H.P.	43'7"	27'7"	306		42-38353	400 @ 25,000
1944	MX-324 Rocket Wing	Aerojet XCAL-200	200 lbs. thrust	32'	12'	244			300 m.p.h.
1944	JB-1 (MX-543)	General Electric	(2) Turbojets	17'6"	10'				400 m.p.h.
1944	JB-10	Ford Pulse Jet	800 lbs. thrust	29'	12'				426 m.p.h.
1943 1944 1945	XP-79 XP-79A XP-79B	Aerojet "Rotojet" " Westingtonhouse 14B	2,000 lbs. thrust " (2) 1345 lbs. thrust	36' 38' 38'	11'4" 11'8" 14'	255 278 288	5:1	43-52437	500+ EST 500+ EST.
1946	XB-35	P&W Wasp Major R-4360-17 & 21	(4) 3,000 H.P.	172'	53'	4,000		42-213603 42-38323	395 m.p.h.
1946	YB-35	P&W Wasp Major R-4360-17 & 21		172'	53'	4,000		42-102366	398 m.p.h.
1947	YB-49	Allison J-35 A-15	(8) 4,000 lbs. thrust	172'	53' 1"	4,000		42-102307 42-102368	520 m.p.h. 30,000
1950	YRB-49A	Allison J-35 A-19	(6) 5,600 lbs. thrust	172'	53'1"	4,000		42-102376	600+ m.p.h.
1950	X-4	Westingtonhouse J-30	(2) 1,600 lbs. thrust	26'10"	23'4"			46-676 46-677	630 m.p.h. @10,000

N1M

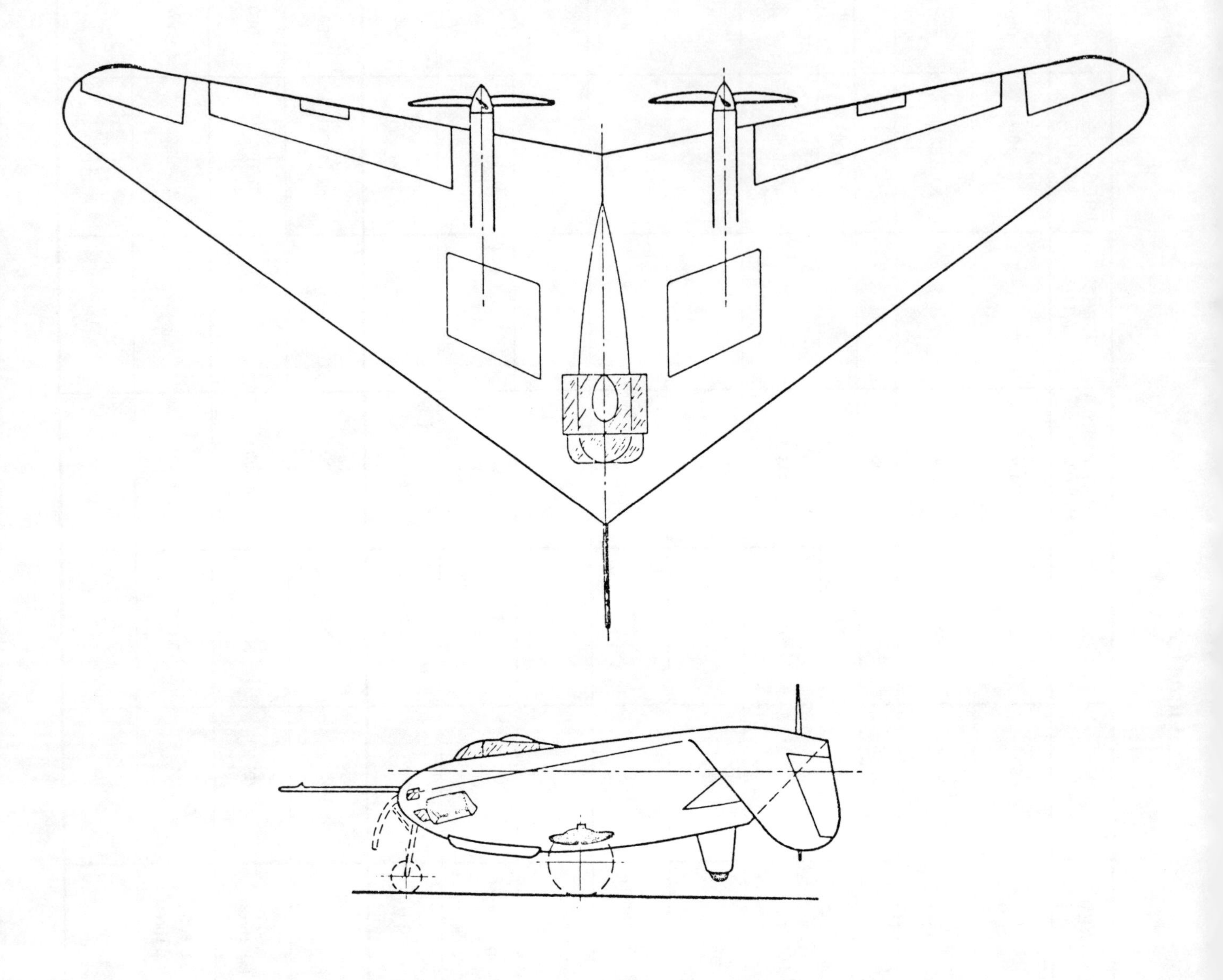

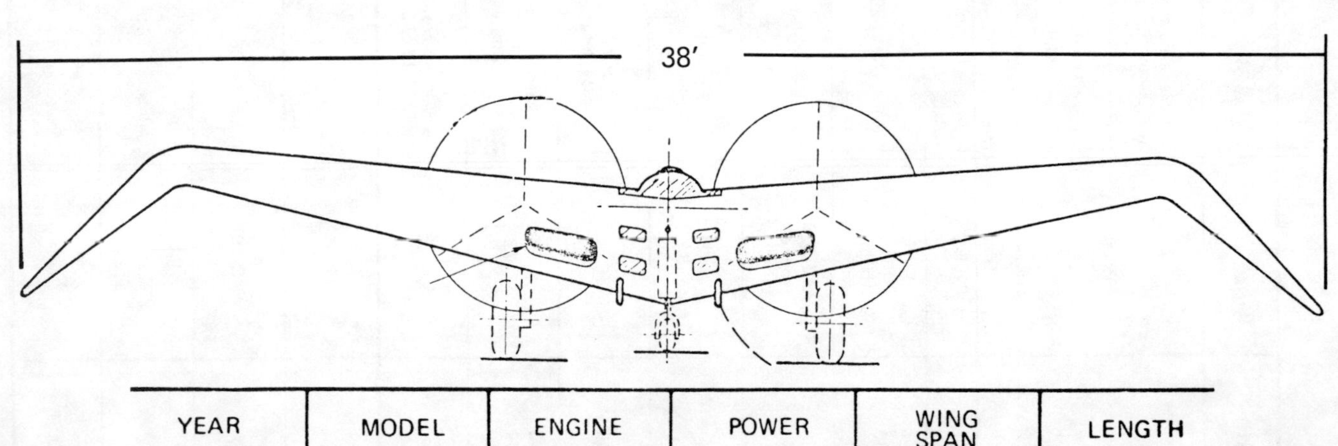

YEAR	MODEL	ENGINE	POWER	WING SPAN	LENGTH
1940	N1M	Lycoming Franklin	(2) 65 H.P. (2) 120 H.P.	38′	20′

MX-324

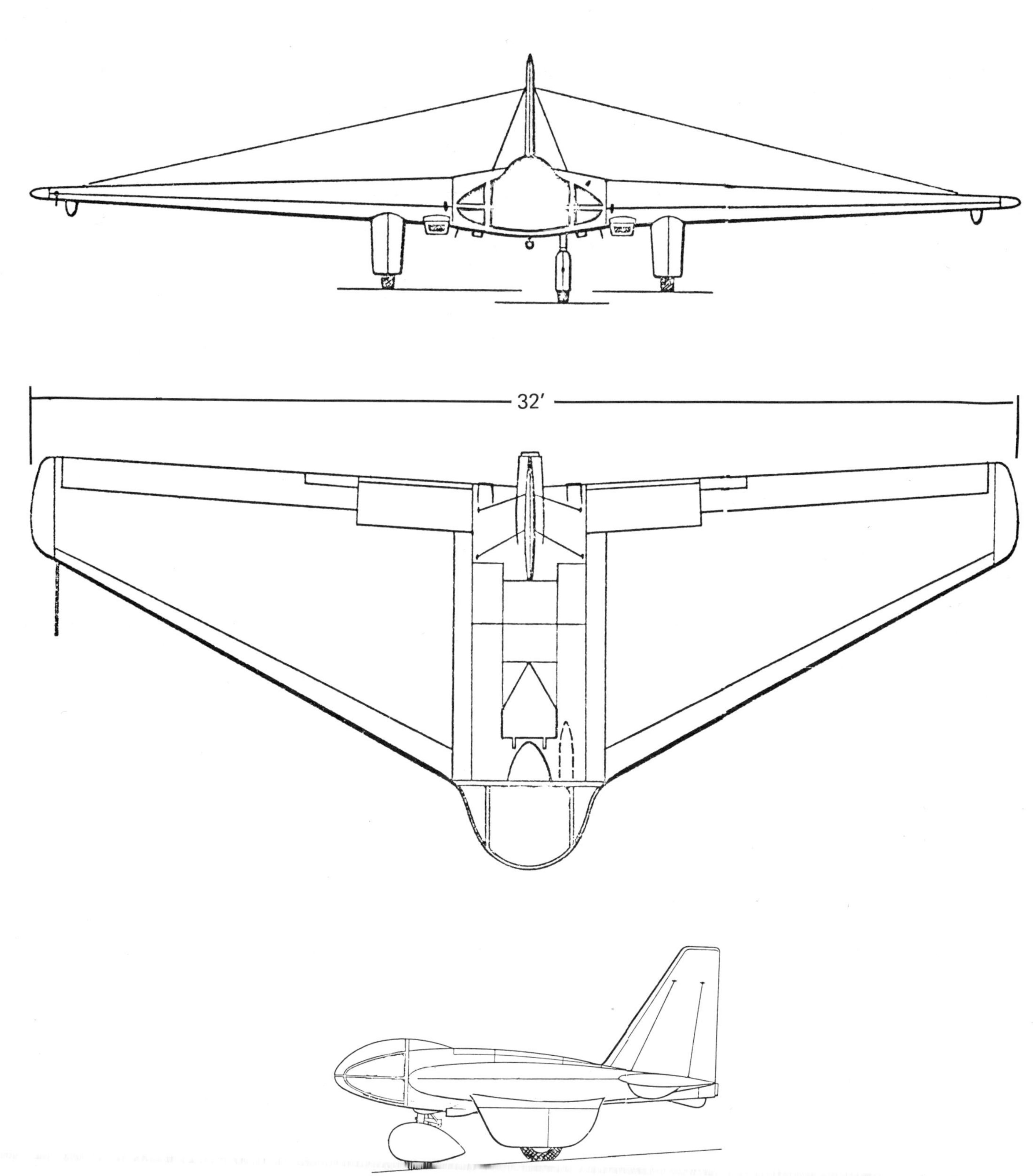

YEAR	MODEL	ENGINE	POWER	WING SPAN	LENGTH
1944	MX-324 Rocket Wing	Aerojet XCAL-200	200 lbs. thrust	32′	12′

XP-79

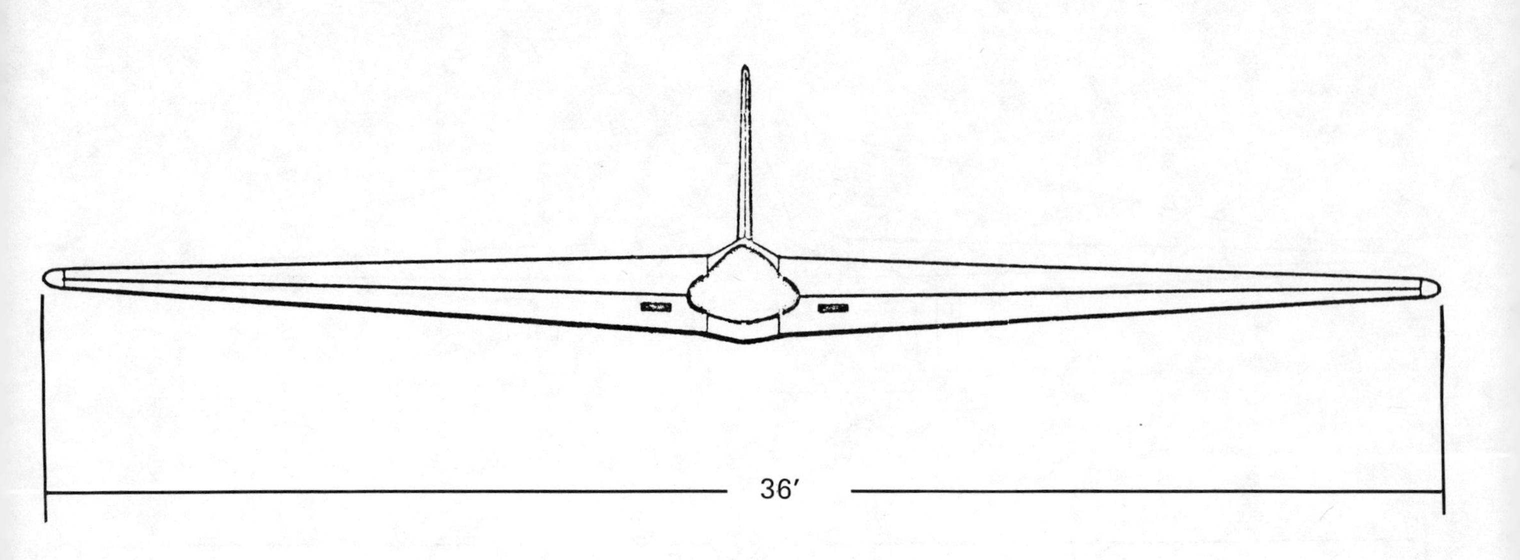

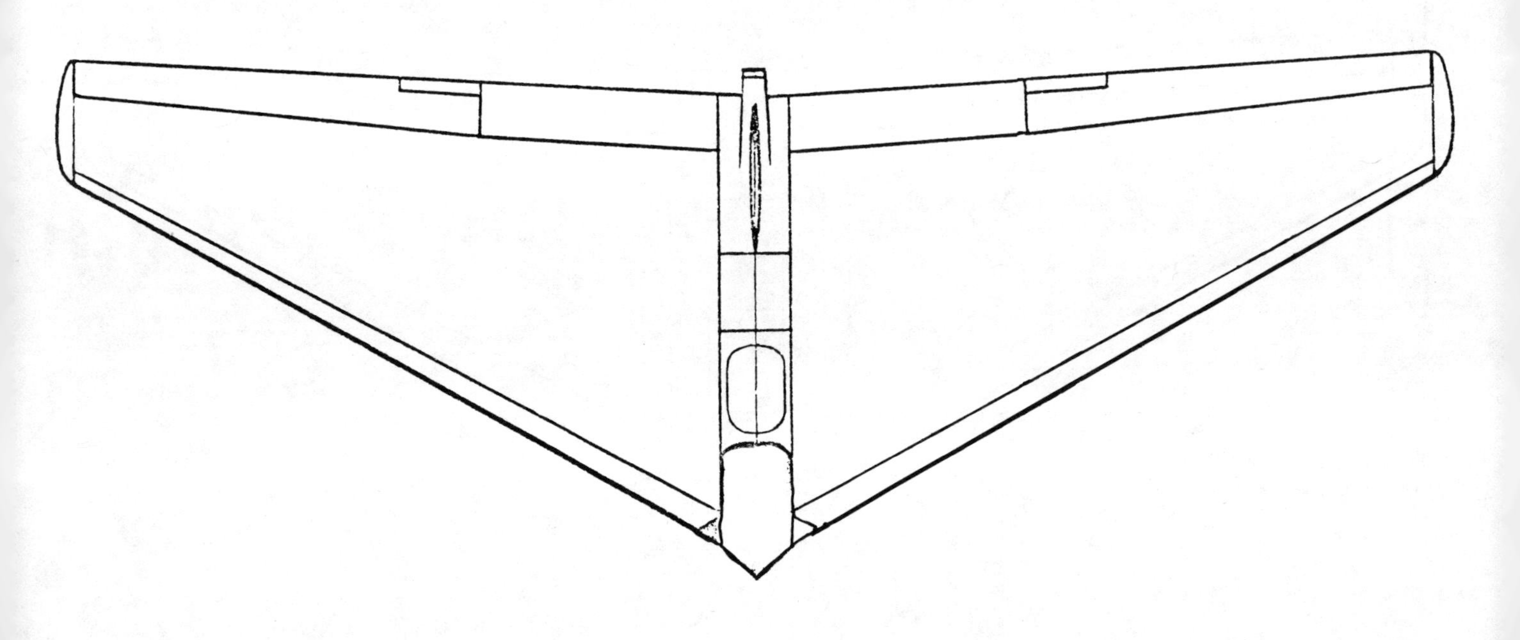

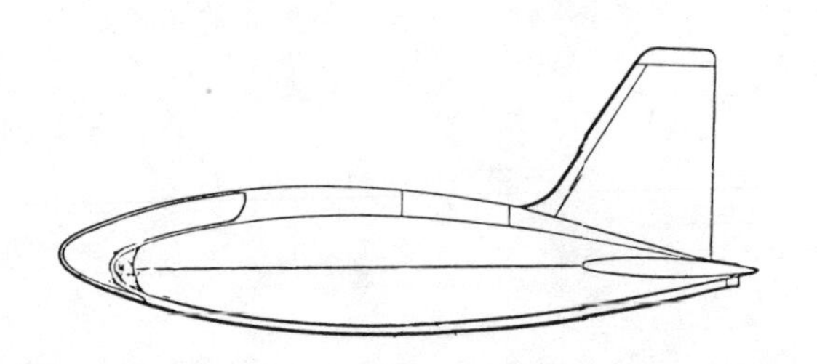

YEAR	MODEL	ENGINE	POWER	WING SPAN	LENGTH
1943	XP-79	Aerojet "Rotojet"	2,000 lbs. thrust	36′	11′4″

XP-79B

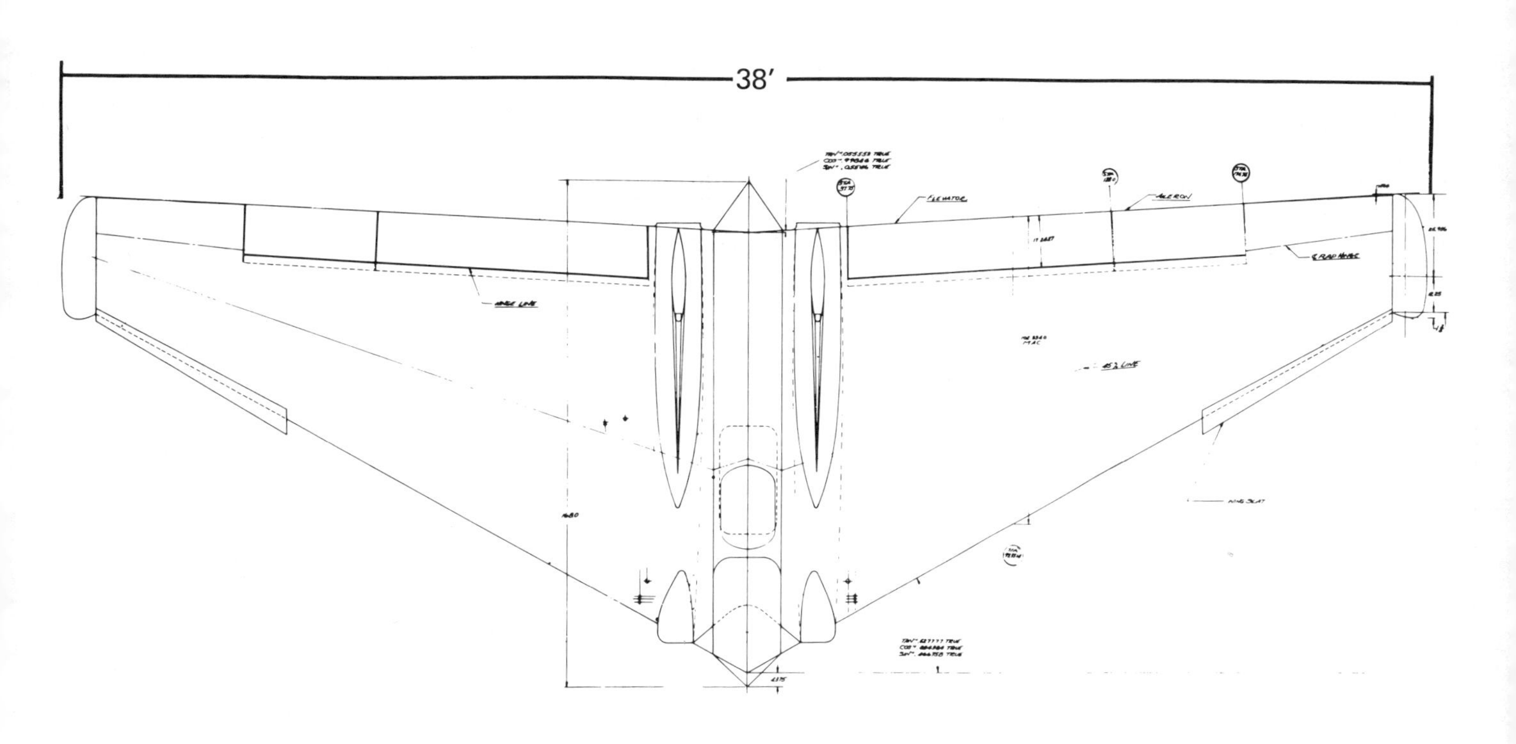

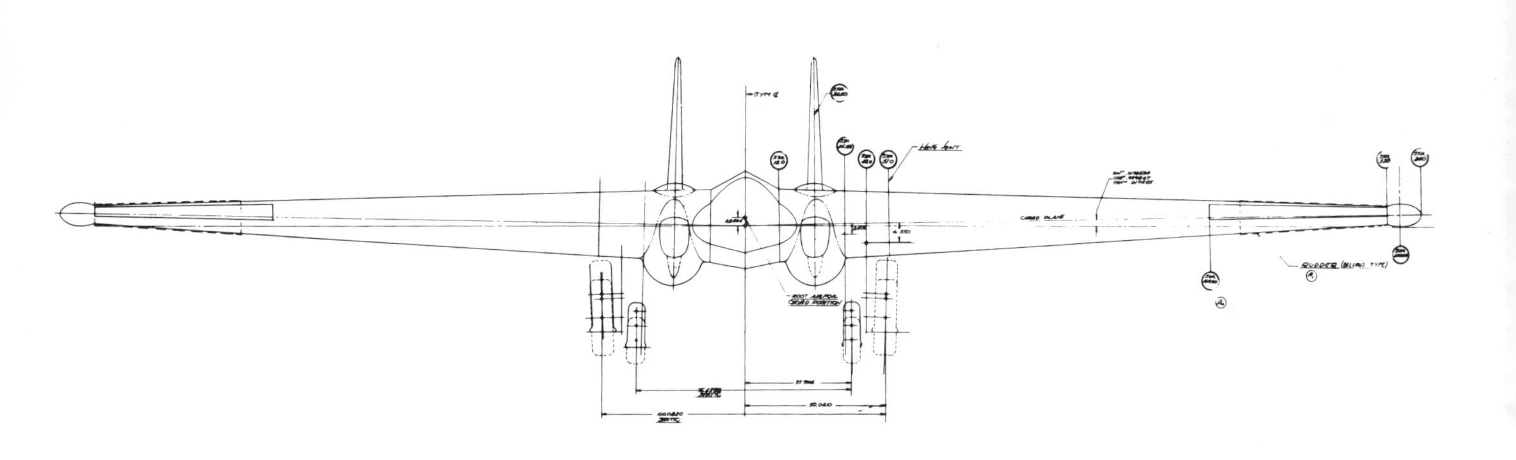

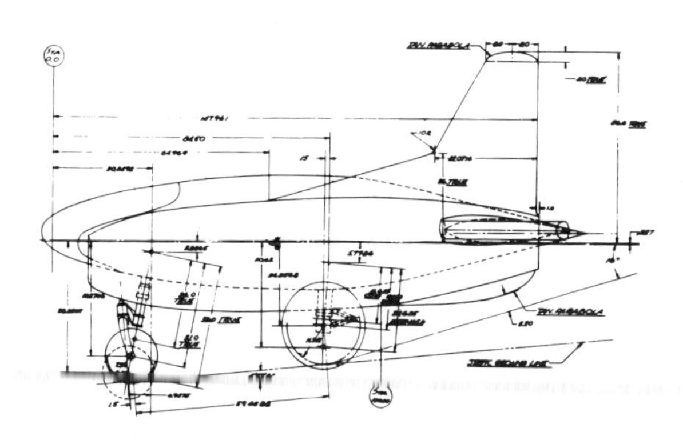

YEAR	MODEL	ENGINE	POWER	WING SPAN	LENGTH
1945	XP-79B	Westingtonhouse 14B	(2) 1345 lbs. thrust	38'	14'

YB-49

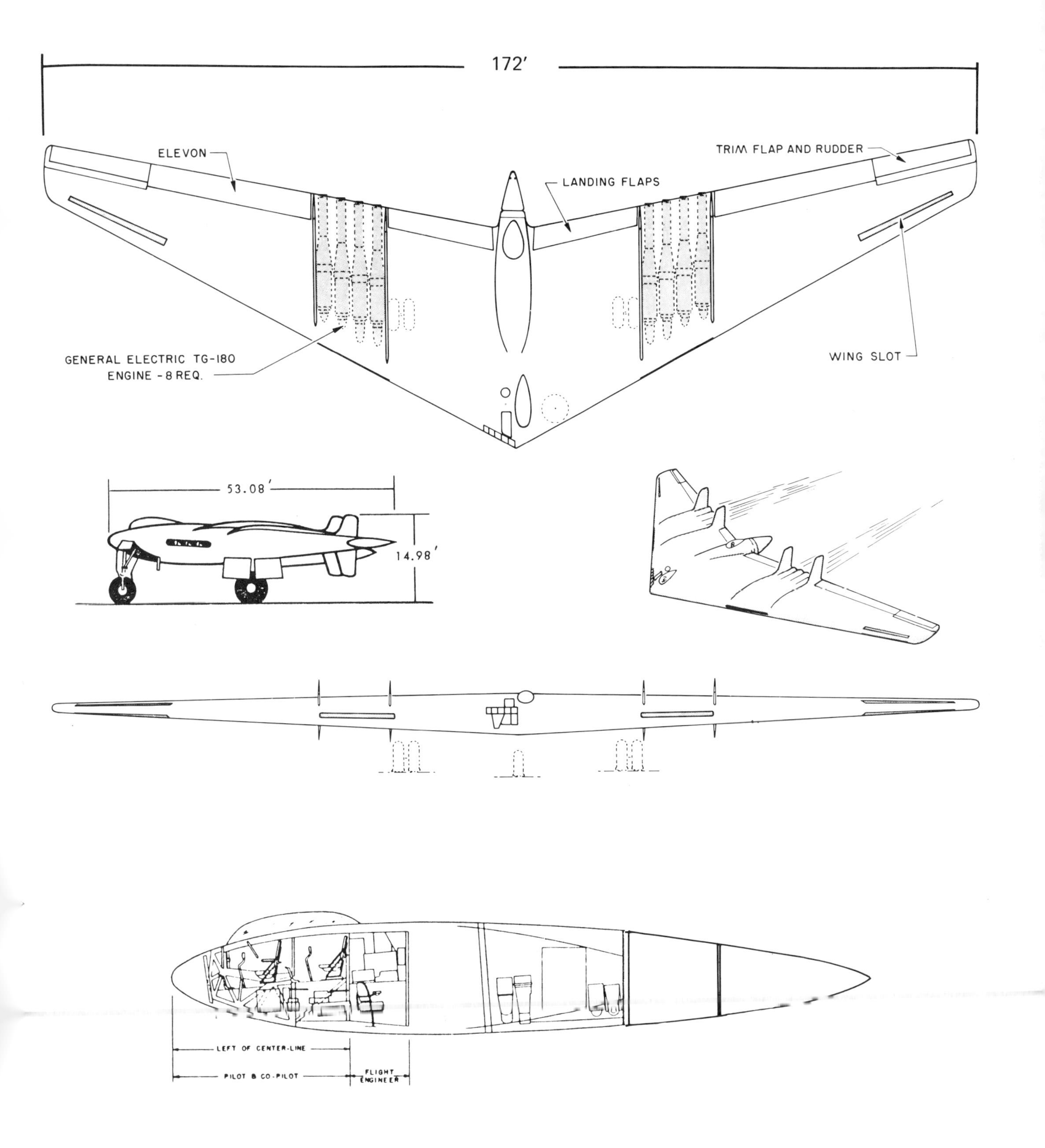
172′
ELEVON
TRIM FLAP AND RUDDER
LANDING FLAPS
GENERAL ELECTRIC TG-180
ENGINE - 8 REQ.
WING SLOT
53.08′
14.98′
LEFT OF CENTER-LINE
PILOT & CO-PILOT
FLIGHT ENGINEER

YRB–49A

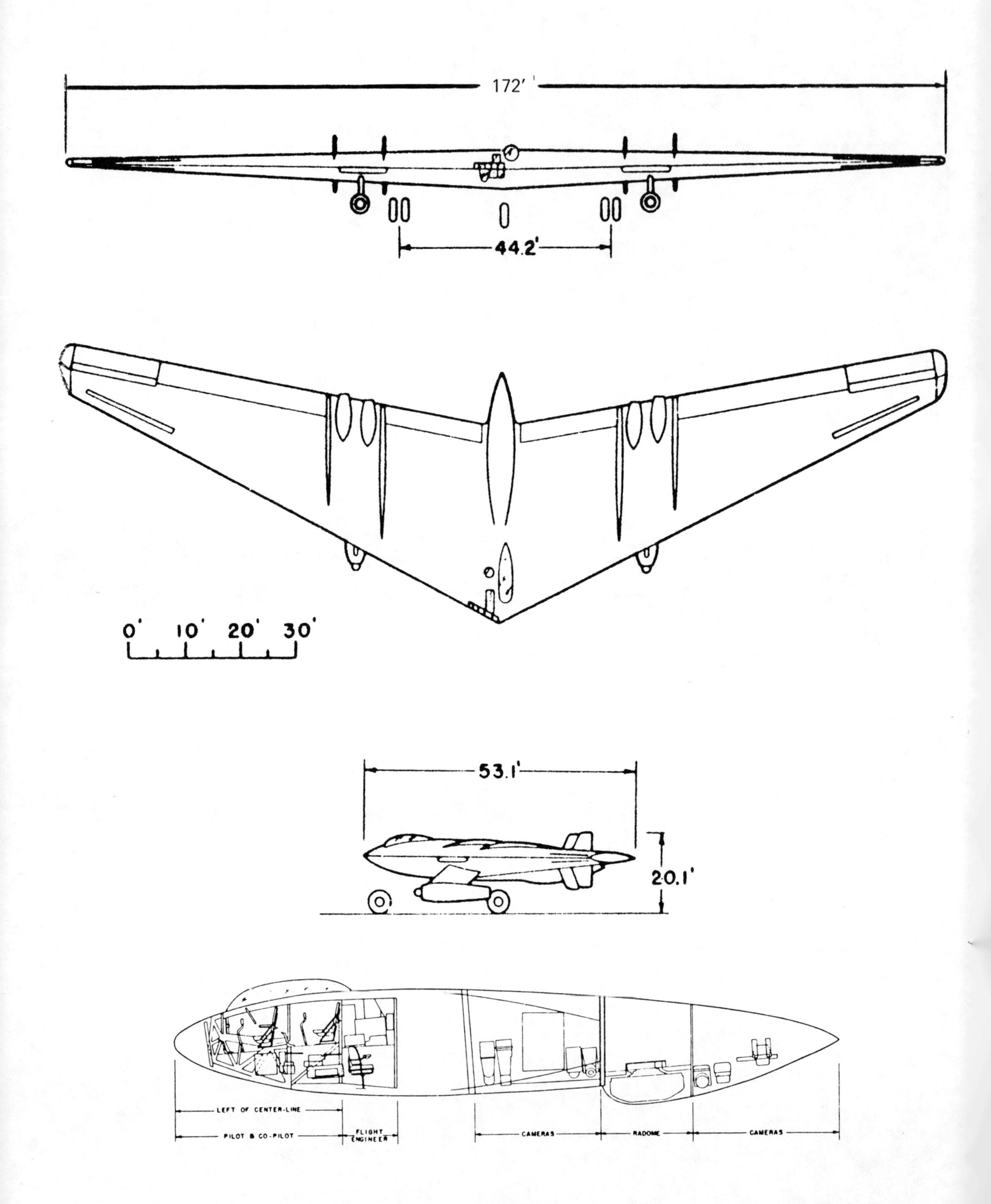

172'
44.2'
0' 10' 20' 30'
53.1'
20.1'
LEFT OF CENTER-LINE
PILOT & CO-PILOT
FLIGHT ENGINEER
CAMERAS
RADOME
CAMERAS

X-4

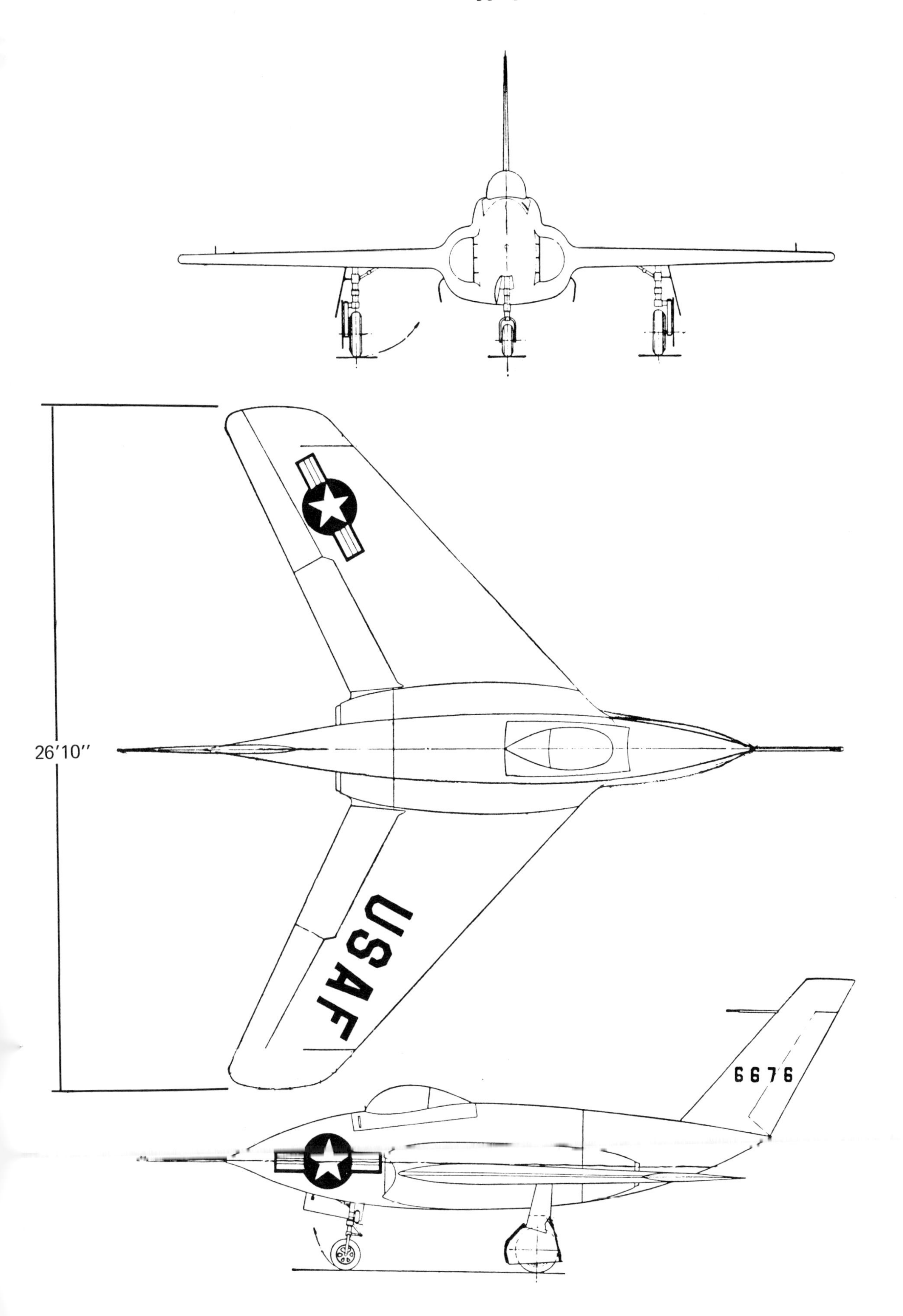

MILITARY SERIAL NUMBER ALLOCATION AND REGISTRATION NUMBERS

	AIR FORCE S/N	C/N	REG. NO.
1929 Semi Flying Wing			NX-216H
N1M			NX-28311
N9M 1		.0036 ?	
N9M 2		.0037 ?	
N9M A		.0038 ?	
N9M B		.0039	
XB-35 #1	42-13603		SCRAPPED AT EDWARDS AFB 1949 (XB-35 through B-35A)
#2	42-38323		
YB-35 #1	42-102366		
#2	42-102369 thru	42-102373	
B-35A	42-102374 thru	42-102378	
YB-49	42-102367		
	42-102368		
YRB-49A	42-102371	N-1496 ?	
EB-35B	42-102375 ?	N-1498 ?	
YB-35B	42-102374 ?	N-1497 ?	
XP-56	42-38353	N-1786	
XP-79B	43-52437		
X-4	46-676*		
	46-677		

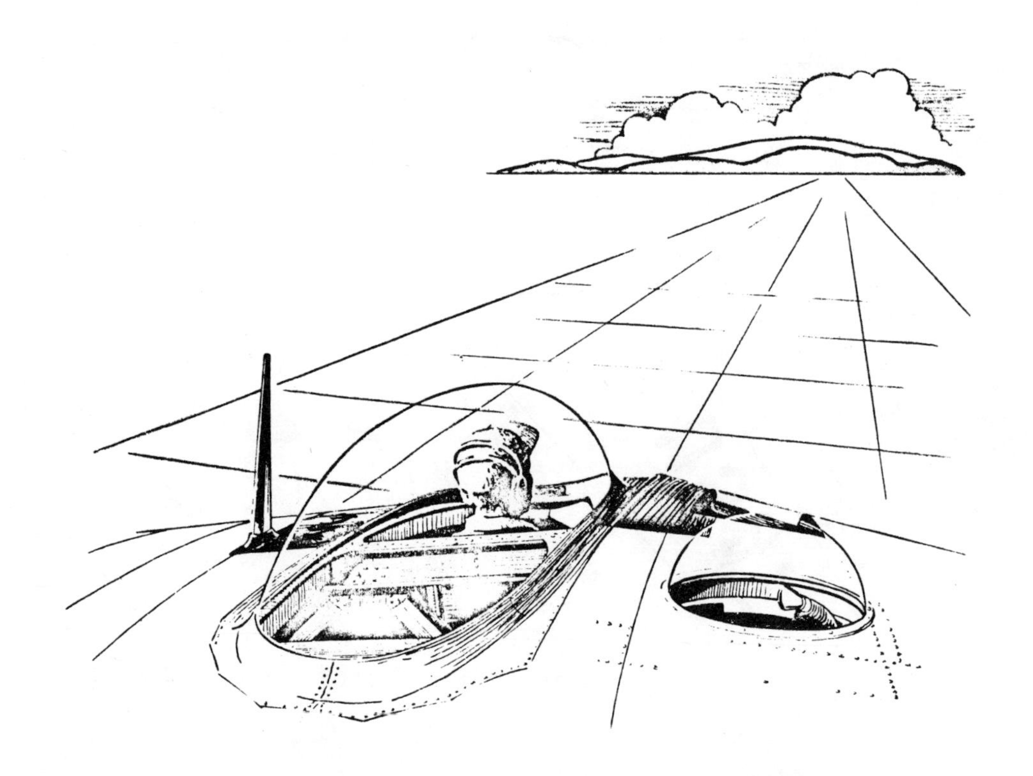

Epilogue

The Northrop Flying Wings are but memories today to those who remember them. This author was privileged to see their sleek wing bodies fly gracefully through the air and to witness the final scrapping of the last YRB-49A.

It is a tragedy that one of the big winged wonder bombers could not have been saved. They were a beautiful sight to behold and a sight that we will probably not see again for some time to come.

As an aircraft, the Northrop Flying Wing was a success. The Wing's major fault was that its design was fifty years ahead of its time.

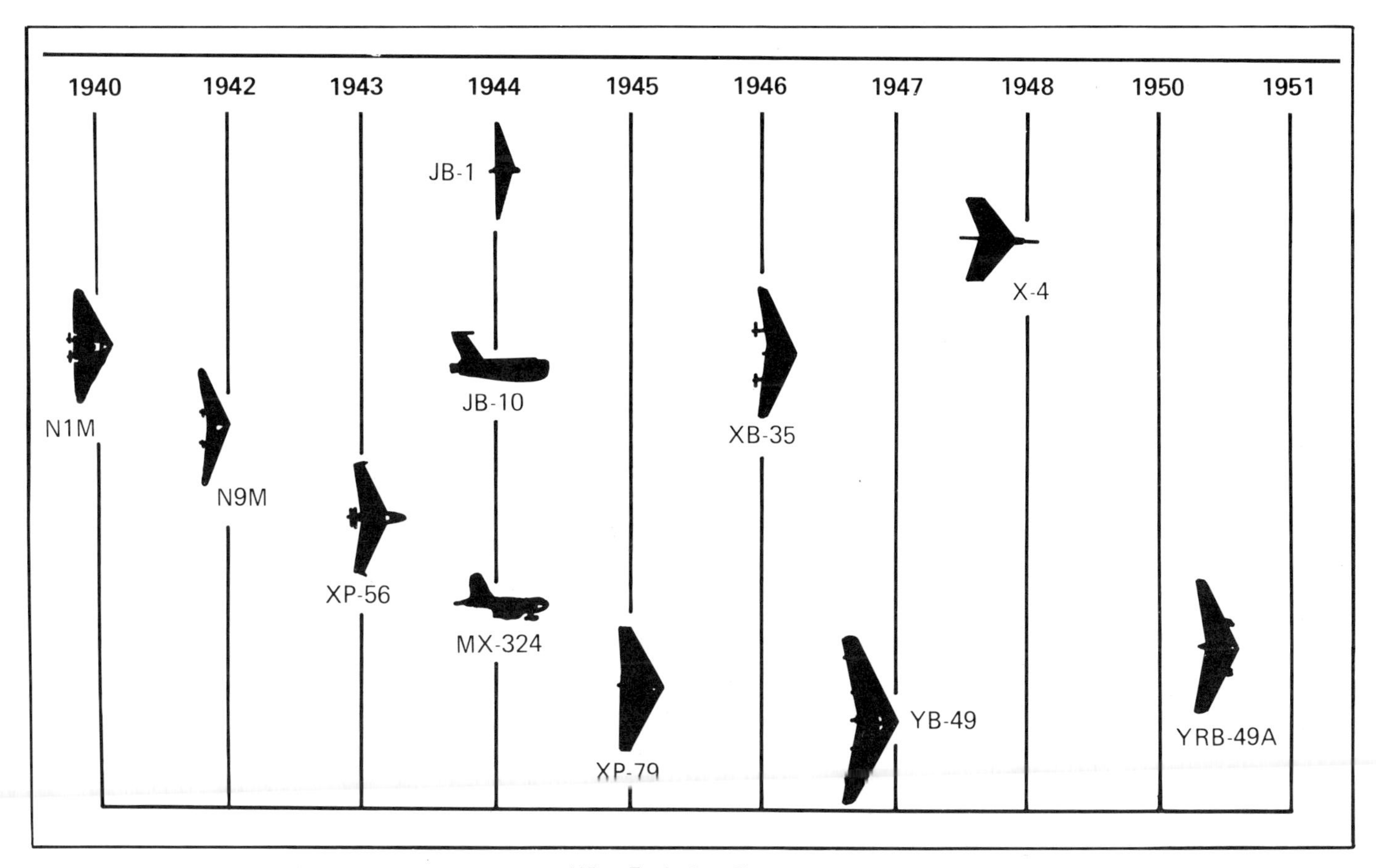

Wing Evolution Chart

. . . Flying the Wing

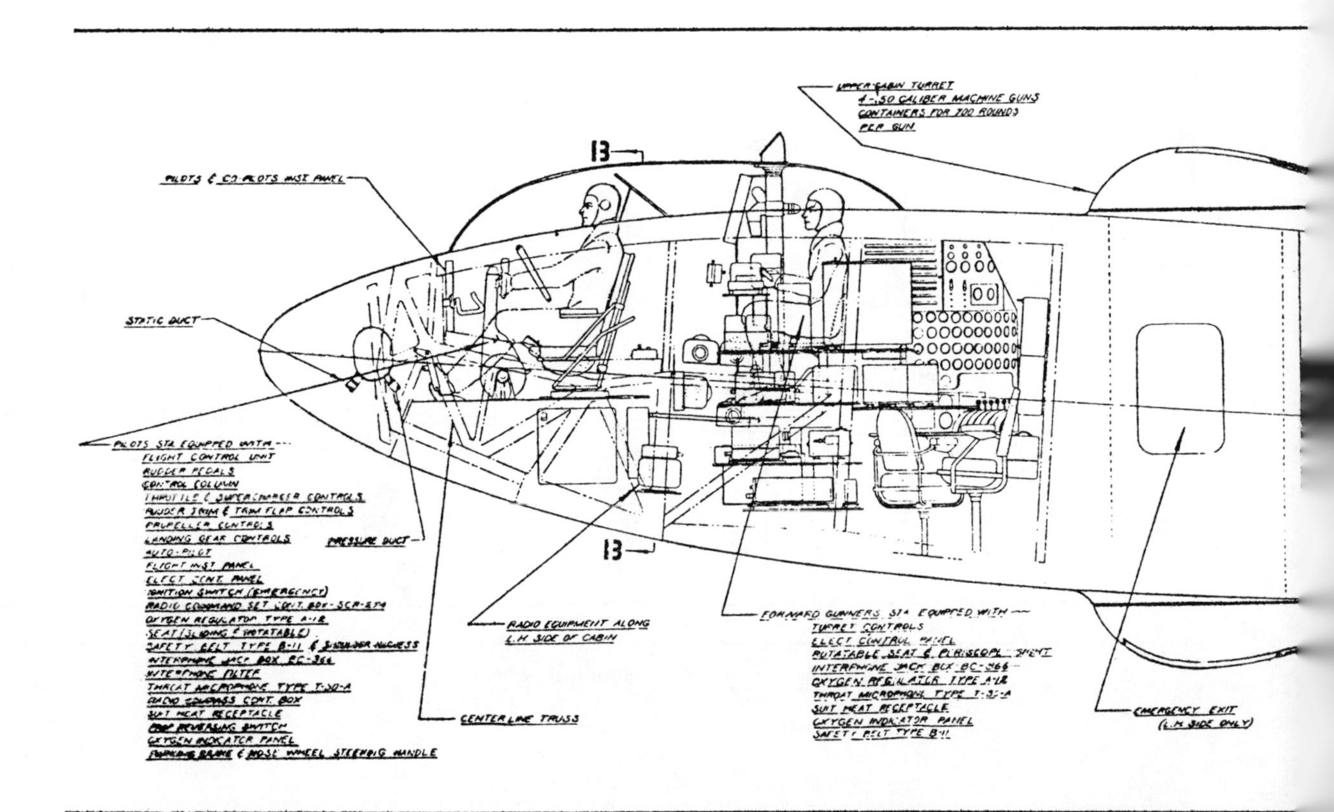

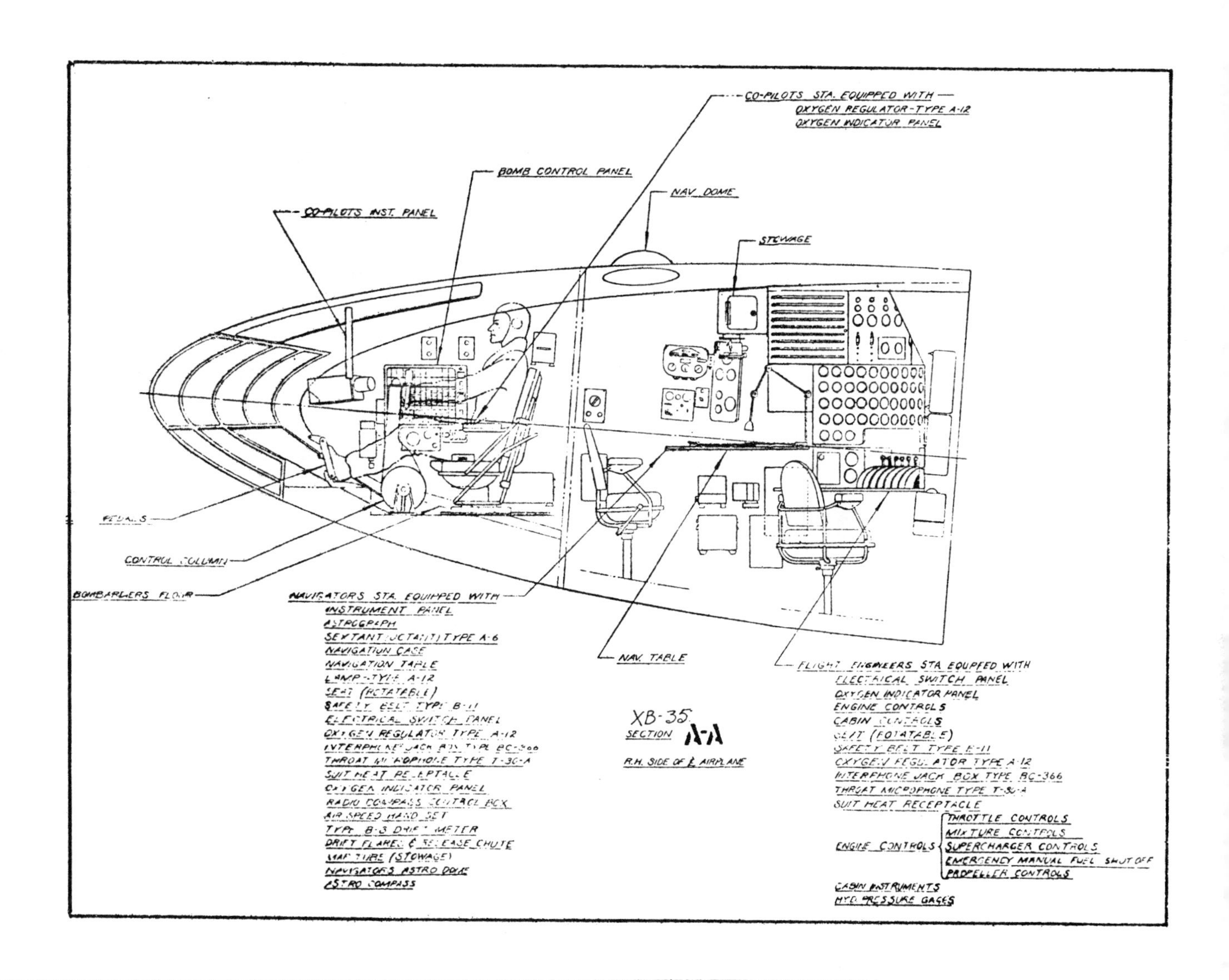
CO-PILOTS STA. EQUIPPED WITH —
OXYGEN REGULATOR - TYPE A-12
OXYGEN INDICATOR PANEL
BOMB CONTROL PANEL
NAV. DOME
CO-PILOTS INST. PANEL
STOWAGE
PEDALS
CONTROL COLUMN
BOMBARDIERS FLOOR
NAVIGATORS STA. EQUIPPED WITH
INSTRUMENT PANEL
ASTROGRAPH
SEXTANT (OCTANT) TYPE A-6
NAVIGATION CASE
NAVIGATION TABLE
LAMP - TYPE A-12
SEAT (ROTATABLE)
SAFETY BELT TYPE B-11
ELECTRICAL SWITCH PANEL
OXYGEN REGULATOR TYPE A-12
INTERPHONE JACK BOX TYPE BC-366
THROAT MICROPHONE TYPE T-30-A
SUIT HEAT RECEPTACLE
OXYGEN INDICATOR PANEL
RADIO COMPASS CONTROL BOX
AIR SPEED HAND SET
TYPE B-3 DRIFT METER
DRIFT FLARES & RELEASE CHUTE
MAP TUBE (STOWAGE)
NAVIGATORS ASTRO DOME
ASTRO COMPASS
NAV. TABLE
XB-35
SECTION A-A
R.H. SIDE OF ℄ AIRPLANE
FLIGHT ENGINEERS STA. EQUIPPED WITH
ELECTRICAL SWITCH PANEL
OXYGEN INDICATOR PANEL
ENGINE CONTROLS
CABIN CONTROLS
SEAT (ROTATABLE)
SAFETY BELT TYPE B-11
OXYGEN REGULATOR TYPE A-12
INTERPHONE JACK BOX TYPE BC-366
THROAT MICROPHONE TYPE T-30-A
SUIT HEAT RECEPTACLE
ENGINE CONTROLS
THROTTLE CONTROLS
MIXTURE CONTROLS
SUPERCHARGER CONTROLS
EMERGENCY MANUAL FUEL SHUT OFF
PROPELLER CONTROLS
CABIN INSTRUMENTS
HYD. PRESSURE GAGES

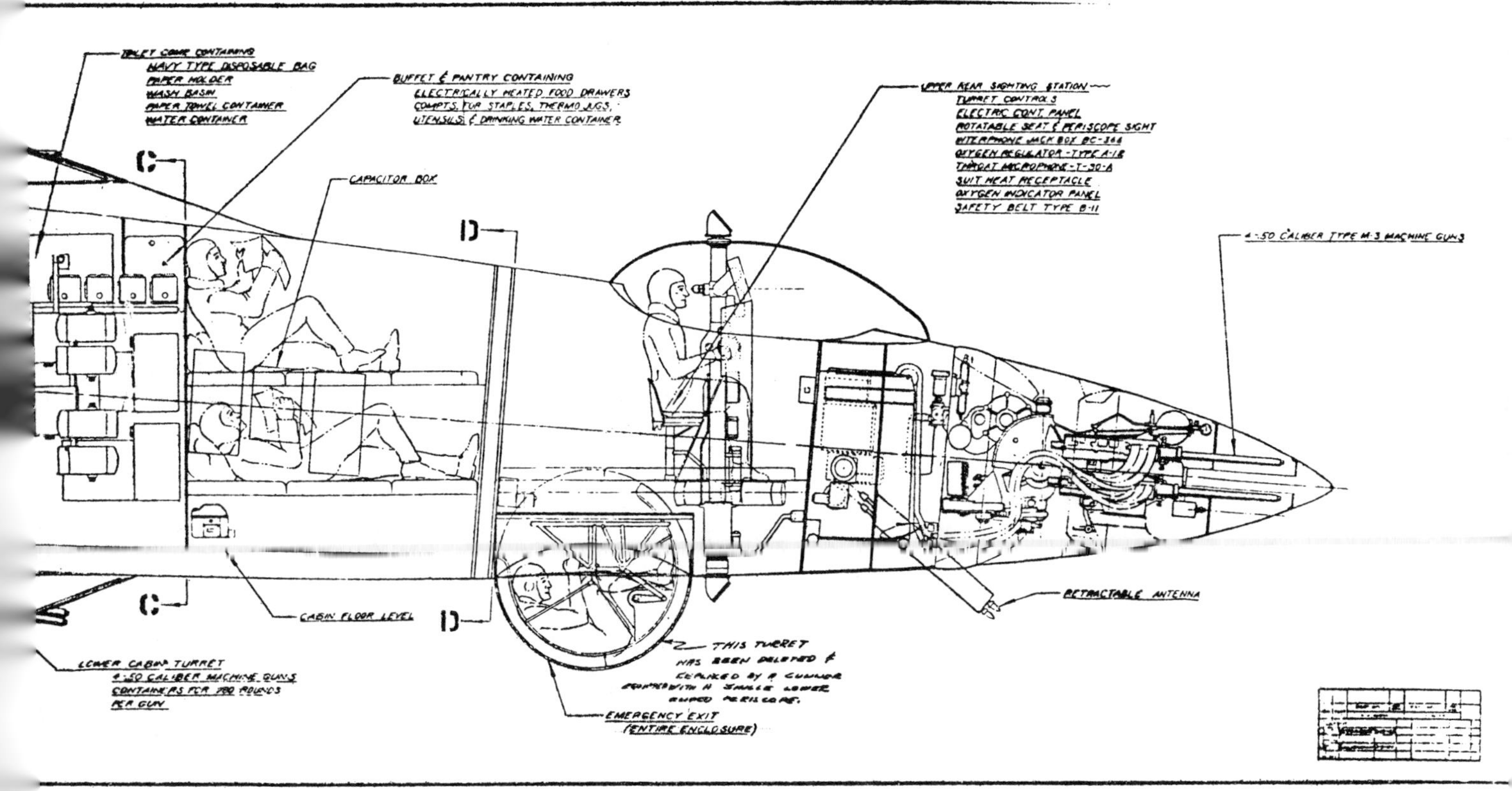
TOILET COMP. CONTAINING
NAVY TYPE DISPOSABLE BAG
PAPER HOLDER
WASH BASIN
PAPER TOWEL CONTAINER
WATER CONTAINER
BUFFET & PANTRY CONTAINING
ELECTRICALLY HEATED FOOD DRAWERS
COMPTS. FOR STAPLES, THERMO JUGS,
UTENSILS & DRINKING WATER CONTAINER
UPPER REAR SIGHTING STATION
TURRET CONTROLS
ELECTRIC CONT. PANEL
ROTATABLE SEAT & PERISCOPE SIGHT
INTERPHONE JACK BOX BC-366
OXYGEN REGULATOR - TYPE A-12
THROAT MICROPHONE - T-30-A
SUIT HEAT RECEPTACLE
OXYGEN INDICATOR PANEL
SAFETY BELT TYPE B-11
CAPACITOR BOX
C
D
4-.50 CALIBER TYPE M-3 MACHINE GUNS
C
D
CABIN FLOOR LEVEL
RETRACTABLE ANTENNA
THIS TURRET HAS BEEN DELETED & REPLACED BY A GUNNER EQUIPPED WITH A SINGLE LOWER PERISCOPE.
LOWER CABIN TURRET
4-.50 CALIBER MACHINE GUNS
CONTAINERS FOR 500 ROUNDS PER GUN
EMERGENCY EXIT
(ENTIRE ENCLOSURE)

YRB-49A EVER SO GRACEFUL.

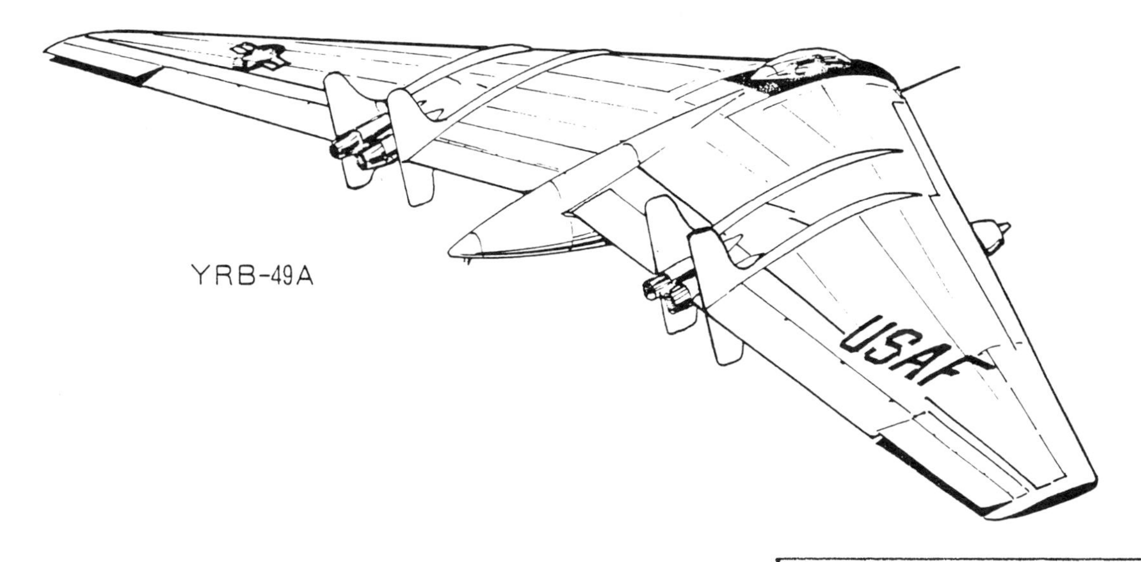
YRB-49A
USAF

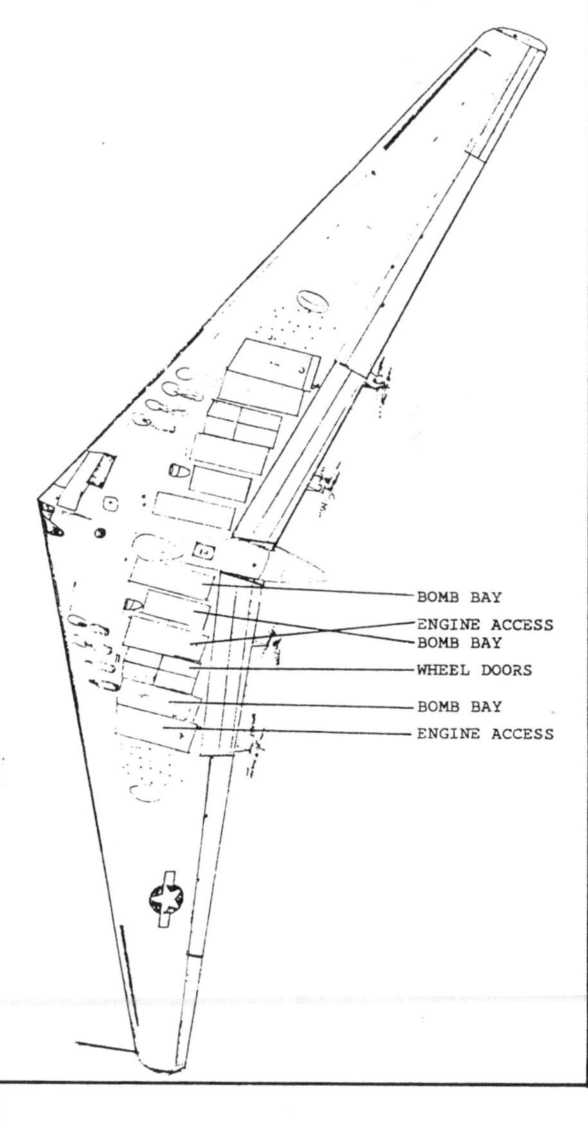
BOMB BAY
ENGINE ACCESS
BOMB BAY
WHEEL DOORS
BOMB BAY
ENGINE ACCESS

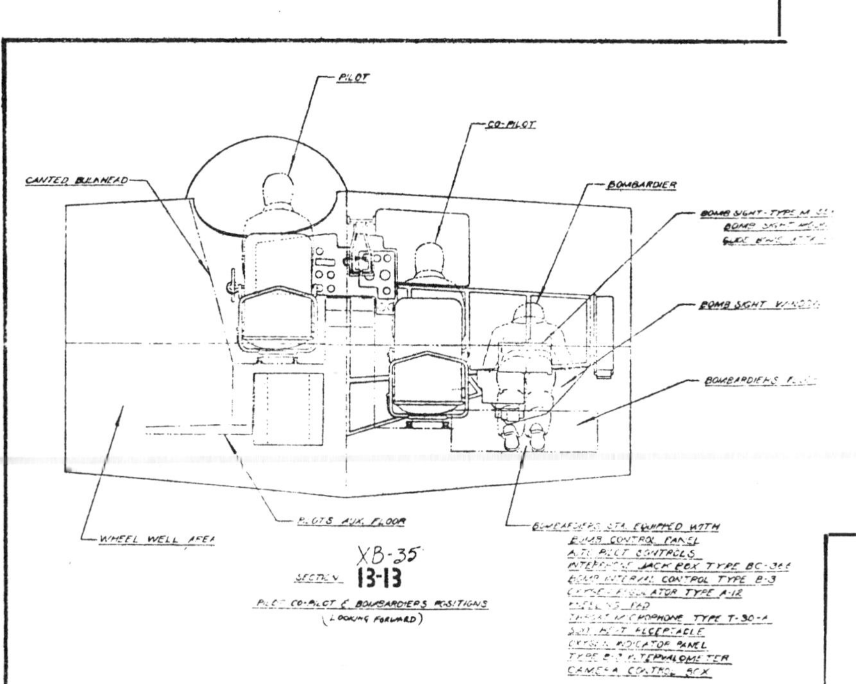
PILOT
CO-PILOT
CANTED BULKHEAD
BOMBARDIER
PILOTS AUX. FLOOR
WHEEL WELL AREA
BOMBARDIERS STA. EQUIPPED WITH
BOMB CONTROL PANEL
AUTO PILOT CONTROLS
THROAT MICROPHONE TYPE T-30-A
CAMERA CONTROL BOX
XB-35
SECTION 13-13
(LOOKING FORWARD)

Says Symington Ordered Merger

Northrop Claims AF Scuttled 'Flying Wing'

By KEN GEPFERT, *Times Staff Writer*

In the years during and immediately after World War II, aviation pioneer John K. Northrop developed and tested an odd-looking, jet-propelled bomber that he believed would revolutionize aircraft design. With no fuselage and no tail, the craft was aptly dubbed the "Flying Wing," and the Air Force selected it to replace the war-tested B-29.

Then, in 1949, Flying Wing production was abruptly canceled and all test planes were ordered destroyed.

For three decades, Northrop has refused to discuss why this promising airplane—the culmination of his lifelong dream—was scrapped so suddenly.

But in a dramatic taped interview broadcast last week, the 85-year-old Northrop Corp. founder finally told his secret: The Flying Wing was canceled, he said, because he refused to obey an Air Force order that he merge his then-fledgling company with a more established competitive firm. When he balked, Northrop said, the Air Force summarily awarded the bomber contract to the competing firm.

Northrop said he kept quiet for all these years because he feared the Pentagon would blackball his company if he disclosed the story. He said he even committed perjury before Congress to hide the facts.

Northrop's allegation shed new light on a generation-old controversy that has become one of the biggest mysteries in American aviation. But it also raised new questions that may never be answered.

In a precise unemotional tone, Northrop told his story to longtime aerospace reporter Clete Roberts in an interview on Los Angeles Public Television station KCET. Since that interview, taped in October, 1979, Northrop has suffered a series of strokes that have left him seriously ill and unable to speak.

The 14-month delay between the interview and its broadcast as part of the KCET documentary last week was due partly to delays in gathering additional material for the telecast and partly to a postponement request by Northrop.

Northrop's story was corroborated by Richard W. Millar, 81, who witnessed the drama as chairman of the Hawthorne-based aerospace

Please see WING, Page 21

WING: Northrop Accuses Symington

Continued from 5th Page

company at the time and who still serves as Northrop vice chairman. But Millar, also interviewed by Roberts, has refused to respond to other questions since the broadcast, saying only that his taped statements "provide an accurate account" of the Flying Wing cancellation.

The Air Force secretary accused of issuing the merger order, former Sen. Stuart Symington (D-Mo.), 79, refused to be interviewed by Roberts. Repeated attempts by The Times to reach both the elder Symington and his son, also a former congressman, were unsuccessful. Most of the other witnesses to events surrounding the Flying Wing cancellation are dead.

Based on the KCET broadcast and subsequent interviews by The Times with Northrop's son and others familiar with the story, however, the picture emerges of a man in a 30-year struggle between his love for the company that bears his name, and for the aircraft that was to be his contribution to aeronautical history.

Northrop, who has long felt his plane had been wronged by history, finally decided to tell his story after becoming convinced—incorrectly as it turned out—that the National Aeronautics & Space Administration was about to resurrect his basic idea.

The Flying Wing bomber was the product of more than 20 years of experimentation by Northrop, who believed as early as 1929 that a plane that was all wing would out-perform traditional designs featuring wings, fusilage and tail assembly. By putting the 15-man crew, eight engines and the bomb bay inside the wing, Northrop minimized the plane's drag and maximized its lift. As a result, the Flying Wing would carry a payload that was nearly equivalent to the plane's weight—a feat matched by no previous aircraft.

Wins Competition

To select a bomber to succeed World War II's B-29s, the Air Force pitted Northrop's Flying Wing, designated the B-35 and later the B-49, against a traditionally configured bomber built by Consolidated Vultee Aircraft Corp. (Convair), which later became a division of General Dynamics Corp. The Flying Wing won a competition against Convair's B-36 in 1948 and the Air Force awarded Northrop a contract to build 35 bombers, with the possibility of ultimately producing 200 to 300 planes.

But Northrop's elation turned into disbelief when he and company chairman Millar were summoned to meet Symington shortly after winning the contract in June, 1948, according to their taped statements.

Noting that his was "a very strange story and perhaps difficult to believe," Northrop told KCET reporter Roberts that Symington launched into a "lengthy diatribe" about how the Air Force did not want to sponsor any new aircraft companies because the Pentagon could not afford to support them with continuing business on declining post-war budgets. Then, Northrop said, Symington demanded that Northrop Corp. merge with Convair.

General Reacts

At that point, Northrop recalled, Brig. Gen. Joseph T. McNarney, commander of the Air Materiel Command and subsequently president of Convair said, "Oh, Mr. Secretary, you don't mean that the way it sounds."

"You're . . .right I do," Symington answered, according to Northrop and Millar.

Northrop and Millar told KCET's Roberts that they then visited Floyd Odlum, head of Atlas Corp., which controlled Convair, to discuss a possible merger. But talks soon ended, Northrop said, because Odlum's demands were "grossly unfair to Northrop."

A few days later, Northrop recalled, Symington telephoned him and said, "I am canceling all of your Flying Wing aircraft."

"I said, 'Oh, Mr. Secretary, why?' "

"He said, 'I've had an adverse report,' and hung up," Northrop recounted. "And that was the last time I talked to him and the last time we could reach him by phone or any other way."

As part of the cancellation, Millar added, the Air Force ordered the destruction of seven Flying Wings then under construction. "Those airplanes were destroyed in front of the employees and everybody who had their heart and soul in it," said Millar, his voice cracking.

After the Air Force canceled the Flying Wing and awarded the contract to the competing Convair B-36, a House Armed Services subcommittee held hearings in 1949 to investigate allegations that the Pentagon used coercion in its aircraft procurement practices.

Prompted by Rumors

According to press accounts at the time, the investigation was prompted by "ugly rumors" about Symington and other Pentagon officials. One rumor investigated—and denied by witnesses at the hearing—was that Symington had been considered to head the firm that would result from the proposed merger between Consolidated Vultee and Northrop.

Among the witnesses who denied seeing any evidence of Pentagon coercion was John K. Northrop.

Northrop testified that he did not "feel there was any unjustifiable or unreasonable pressure in the cancellation of the B-49 contract. I would call the move reasonable and logical." When asked under oath if he was in fear of Pentagon reprisal, Northrop laughed and said, "I have no fear of reprisal."

Thirty-one years later, when asked about his testimony by reporter Roberts, Northrop responded, "My reaction is that under pressure of the life or death of Northrop Corp., I committed one of the finest jobs of perjury that I've ever heard."

Northrop said in the taped interview that he did not tell the full story until now because he feared that Symington would cause the "complete obliteration" of his company. Millar said that the meeting with Symington was so "brutal and bare-faced" that "you almost had to assume that he would be prepared to take further steps if we didn't do as good boys and go along."

After serving as Air Force Secretary, Symington was elected to the U.S. Senate, where he remained for 24 years. He was an influential member of both the Armed Services and Foreign Relations committees, and unsuccessfully ran for the Democratic presidential nomination in 1960. He retired from the Senate in 1977.

Through his secretary in Washington, Symington told reporter Roberts that he "never did (the) sort of thing" alleged by Northrop and Millar.

Prior to Northrop's account, a popular explanation for the Flying Wing's demise was technical failures. The aircraft did exhibit stability and control problems during testing, and one test plane disintegrated during a routine flight in 1948, killing all five crew members.

The Air Force was apparently convinced enough that problems were being corrected to award the production contract to Northrop just five days after the accident, however. The accident investigation proved inconclusive.

Because of Symington's refusal to answer questions and the death of such key witnesses as Gen. McNarney, Convair chief Odlum, and post-war Defense Secretary

Please see WING, Page 23

WING: Northrop Says He Resisted AF Pressure

Continued from 21st Page

Louis Johnson, there may never be consensus on the fate of the Flying Wing.

Whatever the reasons, there is no question Northrop personally was devastated by the cancellation and destruction of the Flying Wing—his lifelong obsession.

In 1952, at the relatively early age of 57, Northrop abruptly retired and divested himself of all interest in the company he founded. "At that time, Jack essentially felt his career was over," said historian William A. Schoneberger, who is writing a book on Northrop's life.

According to his son, John H. Northrop of La Canada, Northrop was particularly troubled by persistent historical accounts that portrayed the Flying Wing as a technical failure in light of its cancellation by the Air Force. The younger Northrop told The Times his father decided he could no longer remain silent. After reading that NASA was considering a Flying Wing design for an advanced, fuel-efficient cargo plane.

Northrop asked to meet with NASA officials in early 1979 to explain his ideas about the design—and to tell why his version was killed 30 years earlier.

"It was a fascinating story," said Gerald Kayten, deputy director of NASA's aeronautical systems divisions, who attended the half-day meeting at Northrop University in Inglewood. "But there really wasn't much of a meeting of minds. All Mr. Northrop seemed to be interested in was pointing out that his airplane was a pretty good airplane. He didn't need to convince us, because we already agreed with him." NASA already decided, however, to put the Flying Wing design "on a back burner" because it was best suited for much larger cargo planes than will be needed by military or commercial users for the next two decades, Kayten told The Times.

Nonetheless, in a letter sent to Northrop after the meeting, NASA Administrator Robert A. Frosch acknowledged Northrop's pioneering work and said "our analyses confirmed your much earlier conviction as to the load-carrying and efficiency advantages of this design approach."

Armed with this evidence that the wisdom of his approach finally was being recognized by the government, Northrop asked the company's present management for permission to tell his story, according to Northrop's son.

Even after telling his story to Roberts, Northrop had second thoughts and asked the reporter to delay broadcasting the interview for several months, according to the KCET reporter. "Then one day he called and said, 'Go ahead, Clete. It's all clear now,' " Roberts recalled.

Northrop, now seriously ill in a Glendale hospital, was given a private screening of the documentary before it was broadcast. He could not speak to give his reaction, his son said, but "he put his hands together and shook them, like a fighter does, to show us he was pleased."

STUDIES BY NORTHROP ENGINEERS have disclosed that the famous Flying Wing B-35 and B-49 bombers are admirably adaptable for use as commercial passenger and cargo transports. Exploration has been made of the commercial aspects of Flying Wings and Northrop engineers are convinced passenger-cargo versions are logical and practical applications of the Flying Wing principle. Passengers would be seated in spacious drawing rooms that would provide maximum visibility through transparent leading edges. In flight passengers would be able to view the terrain unfolding ahead and far below. Crew members would be housed in an upper deck faired into a verticle air separator fin. Gone would be the claustrophobic "shut in" feeling common to most public conveyances. Meals would be served on large trays attached magnetically to the arm of the passenger's chair. Northrop transport Wing will operate in 400 to 500 miles per hour speed range. Leaving Los Angeles in the evening, the traveler would consume a leisurely dinner. Afterward, as the Flying Wing flew eastward through the high skies, the latest film program would be shown. When the film was over, Chicago would have been passed, New York would be minutes away. This conception of the interior of a Flying Wing "airliner of tomorrow"

80-Passenger Transport

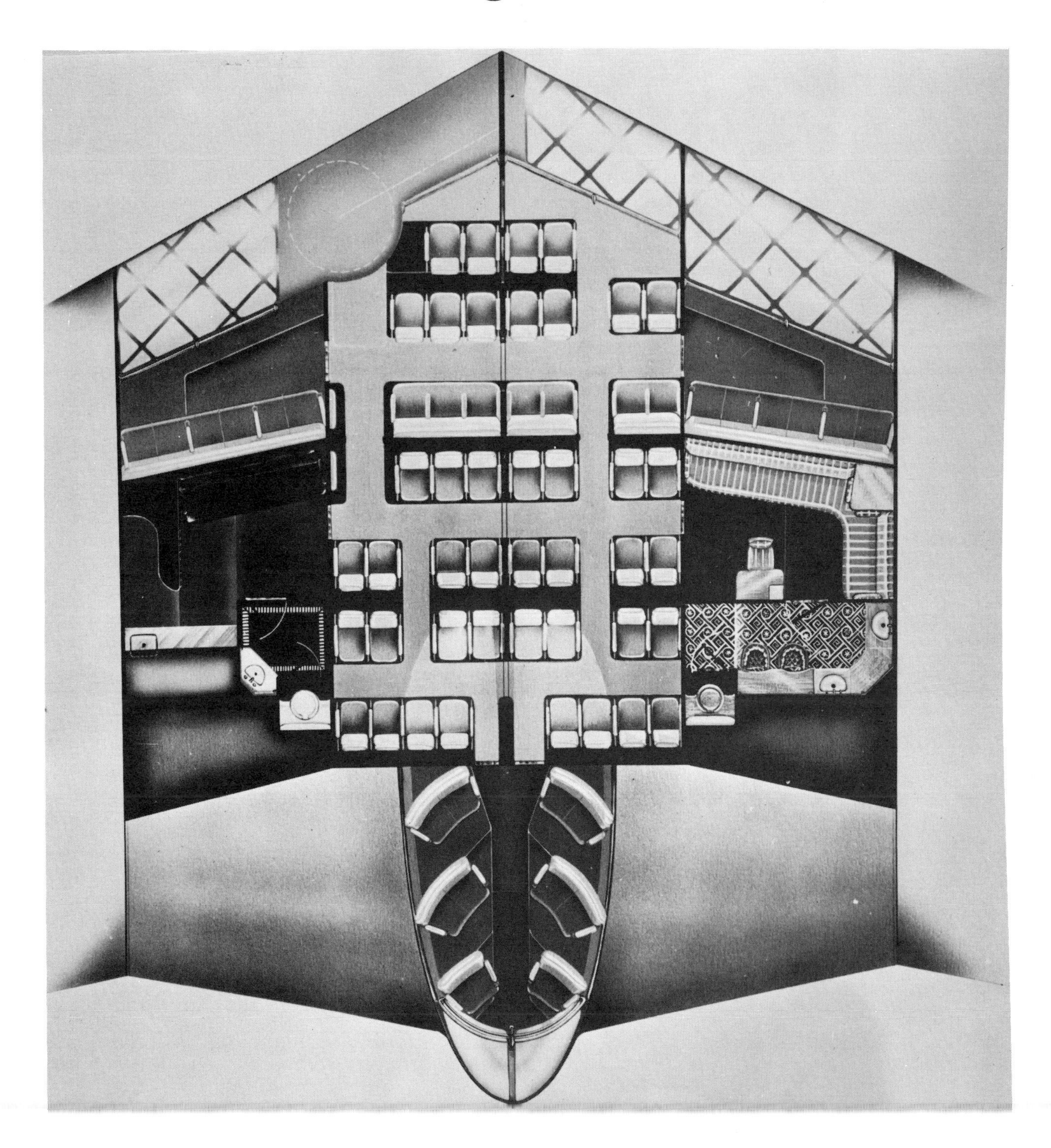

FLYING WING TRANSPORTS

It was anticipated that a transport version would be built following completion of the Flying Wing Bomber program for the U.S. Air Force.

THE OMINOUS STINGER OF THE GIANT Northrop Flying Wing bombers is shown transformed via artist's conception into a transparent stratospheric "observation platform" in the Flying Wing "airliner of tomorrow". Passengers, traveling at greater speeds and lower costs, enjoy at the same time the utmost in modern conveniences. Spacious accommodations depart from the time-honored "stage coach, railroad car" seating arrangements. Passenger quarters are more in the manner of theater seating. Staterooms would be compact editions of those encountered on transatlantic liners. Total comfort throughout for trans-ocean and transcontinent travelers would be Flying Wing liner keynote.

NOTES

NOTES

WORLD WAR II PUBLICATIONS

- **FIGHTER TACTICS OF THE ACES**

By Combat Aces Of South Pacific

For the first time gathered together in one book the fighter tactics employed by the top South Pacific Aces: Bong, McGuire Lynch, Kearby, MacDonald and many many others. Color artwork covers. Aircraft profile drawings. Big Size 8½ x 11 Over 100 photographs, This is the story of how our Air Force won its aerial victories and how our fighter pilots employed superior fighter tactics to gain the advantage over the Imperial Japanese Army and Navy Air Forces in the skies of the South Pacific.

- **"LUFTWAFFE AIRCRAFT & ACES"** **By Edward T. Maloney**

A definitive survey of leading Luftwaffe aircraft – engines, including Jet and Rocket Propulsion, bombs, bombsights, machine guns, cannon, and cockpit details. A special section is included on the leading German Aces. *A Big Book Value.* Over 200 Pictures – 4 Pages of Color Drawings – 152 Pages.

- **"INVASION D-DAY"**

D-Day June 6th, 1944, as viewed in photographic detail by the other side. A German account of the Battle & forces under the German High Command on this famous day in history. Size 7 x 9¾ – 55 Pages – 103 Photos.

- **"NORTHROP FLYING WINGS"**

By Edward T. Maloney

Far reaching coverage of sixteen different Flying Wing designs of John Northrop, dating from 1928 to the super secret jet powered YRB-49A and the X-4 Bantam Wing. Contains wing evolution chart, specifications, flight control operations. Over seventy excellent, and rare photographs. Three view drawings including two fold out factory drawings.

- **"SEVER THE SKY"** **EVOLUTION OF SEVERSKY AIRCRAFT**

Amazing story and development of one of America's greatest fighter aircraft and of its founder Maj. Alexander P.de Seversky and his designer Alex Kartveli. Its Air Racing History and Hollywood Film use are covered as well as its wartime exploits. Many never before published photo's are also included. Complete 1/72 Scale 3-View Drawings, Specifications, over 130 pages, soft covers, Big Size 8x11

- THE MESSERSCHMITT Me 262 over 200 photos, six pages of scale drawings covers all models: trainers, day fighters, night fighters, and intercepter versions, advanced designs and lists surviving aircraft....Me 262 in combat etc. Size 8½ x 11. Softbound

- **ZERO FIGHTER OVER JAPAN**

History of the famous Japanese Navy WWII.,Zero Fighter, the story and history of the Zero Model 52 restore by Planes of Fame Museum, Chino Airport, Calif. U.S. Test Pilots who flew it during WWII. Test pilot report by Capt. Don Lykins. Color cover, 32 pages on high gloss quality papers, 33 photographs, soft covers, size 7¼ x 10¼

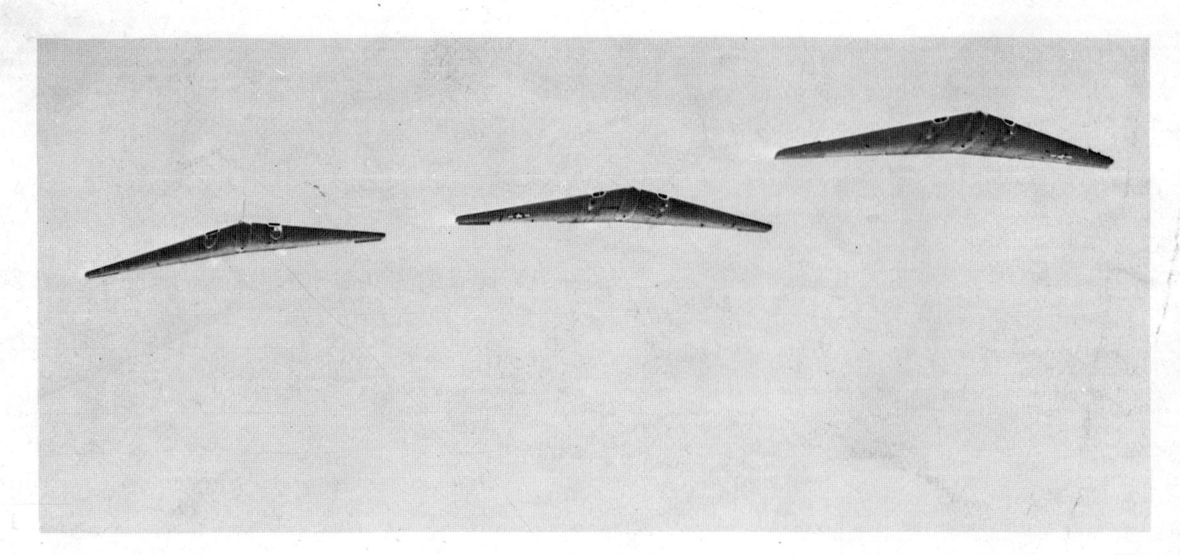